NEUER BOTANISCHER GARTEN

SHANGHAI

NEW BOTANIC GARDEN

jovis

NEUER BOTANISCHER GARTEN

SHANGHAI

NEW BOTANIC GARDEN

DONATA + CHRISTOPH VALENTIEN

jovis

INHALT CONTENT

VORWORT

HE SHANAN

Weltweit blicken wir auf eine Geschichte der botanischen Gärten von mehreren 1000 Jahren zurück. Nun beginnt mit dem Neuen Botanischen Garten Shanghai ein weiteres Kapitel. Nach den exotischen Pflanzen und der Pflanzensystematik steht heute die Pflanze im Ökosystem im Vordergrund. Es entsteht ein botanischer Garten, der Erholung und Bildung mit der Tradition der Region und der Verbesserung der Lebensqualität verbindet.

In der Gartenwissenschaft ist der botanische Garten ein spezieller Zweig, der aus der Wissbegierde der Menschen und dem Bedarf an der Nutzung der Pflanzen heraus entstand. Botanische Gärten sind aus dem europäischen Kulturkreis hervorgegangen. Die botanischen Gärten der Universitäten in Italien aus der Mitte des 16. Jahrhunderts und die Gärten in Pisa und Padua sind die ersten anerkannten botanischen Gärten der Welt. Seit es den botanischen Garten gibt, wird in ihm wissenschaftliche Grundlagenforschung zur Schau gestellt. Die Entwicklung dieser Gärten und damit der Botanik, insbesondere der Pflanzensystematik, haben eine enge Verbindung zur sozioökonomischen Entwicklung der Menschheit.

Die Entwicklung der botanischen Gärten in den vergangenen 500 Jahren lässt sich in drei Phasen aufteilen. Die erste Phase umfasst den Zeitraum von der Mitte des 16. Jahrhunderts bis zur Mitte des 19. Jahrhunderts. Damals gab es weltweit etwa 100 botanische Gärten. Infolge der binären Nomenklatur des schwedischen Forschers Carl von Linné aus dem 18. Jahrhundert vervielfachte sich das menschliche Wissen um den Reichtum an Pflanzenarten; dies schuf die Grundlage für die Entwicklung der botanischen Gärten.

Mit dem Anstieg der gesellschaftlichen Produktivität und der Anerkennung der Darwin'schen Evolutionslehre nach der Mitte des 19. Jahrhunderts begann die zweite Phase der Entwicklung der botanischen Gärten. Diese zweite Phase, die bis zur Mitte des 20. Jahrhunderts andauerte, war weltweit die fruchtbarste und wichtigste Phase in der Entwicklungsgeschichte der botanischen Gärten. Die Entwicklung und Nutzung der wirtschaftlich bedeutsamen Pflanzen, mit dem Rohstoff Gummi als Beispiel, wurde in allen botanischen Gärten mit Begeisterung vorangetrieben. Die Anzahl der botanischen Gärten in den britischen Kolonien stieg von acht auf über 100 und bis zum Zweiten Weltkrieg waren es weltweit bald über 500.

Nach dem Ende des Zweiten Weltkriegs, vor allem, als der menschlichen Gesellschaft die globale ökologische Krise nach und nach bewusst wurde, wurde angesichts der sich schnell ändernden Welt der Schutz der Pflanzen zur neuen Aufgabe der botanischen Gärten – eine Herausforderung, die zu einer neuen Entwicklungschance der botanischen Gärten führte. Die Länder der Dritten Welt, die reichhaltige Pflanzenressourcen besaßen, hießen die stetige Entwicklung der botanischen Gärten willkommen. Nach und nach wurden botanische Gärten eingerichtet, sodass ihre Anzahl auf der Welt erneut rasch anstieg. In dieser Phase verlagerten sich die Schwerpunkte in den Industriestaaten in Europa und in den Vereinigten Staaten zur Grundlagenforschung, Bildung und zum Schutz der von globalen ökologischen Problemen bedrohten Pflanzenarten.

Die nachhaltige Entwicklung der Gesellschaft mit dem Ziel eines harmonischen Miteinanders von Mensch und Natur führte dazu, dass die Aufgaben der botanischen Gärten immer umfassender wurden. Es wurde zu einer unbestreitbaren Tatsache, dass Städte mehr botanische Gärten brauchen. Bis heute gibt es weltweit mehr als 2400 botanische Gärten, die insge-

FOREWORD

HE SHANAN

The history of botanic gardens around the world stretches back several thousand years. With the new Shanghai Botanic Garden a further chapter has begun. In addition to exotic species and the classification of plants, attention is now also focused on vegetation as part of the ecosystem. A botanic garden is being created which will combine recreation and education with the region's tradition and the improvement of the quality of life.

Botanic gardens represent a special branch within the science of gardens which originated from people's thirst for knowledge and from the need to use plants. This type of garden originally developed from within the European cultural sphere. The botanic gardens of Italian universities in the middle of the 16th century and the gardens in Pisa and Padua are the first recognised botanic gardens in the world. Fundamental scientific research has been put on exhibit since the very beginning of these gardens. The development of botanic gardens, and thus of botany, especially the classification of plants, have a close connection to the socio-economic development of humanity.

The development of botanic gardens in the past 500 years can be divided into three phases. The first phase took place between the middle of the 16th century and the middle of the 19th century. During this period there were approximately 100 botanic gardens worldwide. As a result of Swedish researcher Carl von Linné's binary nomenclature in the 18th century, human knowledge about the wealth of plant species multiplied, which in turn created the basis for the development of botanic gardens.

The second phase of the development of botanic gardens began with the increase in social productivity and the recognition of Darwin's theory of evolution after the middle of the 19th century. This phase lasted until the middle of the 20th century and was the most fruitful and important phase worldwide concerning the development of botanical gardens. The development and use of economically important plants, the raw material rubber, for example, was enthusiastically promoted in all botanic gardens. The number of Royal Botanic Gardens in the British colonies increased from eight to over 100. In only 100 years, until the Second World War, the number of botanic gardens around the world increased quickly from over 100 to over 500.

After the end of the Second World War, and especially when human society slowly became conscious of the global ecological crisis, plant protection became a new task for botanic gardens, particularly in view of the rapidly changing world. This challenge led to a new opportunity for development. Third world countries with extensive vegetative resources welcomed the steady development of botanic gardens. Botanic gardens were gradually built, and their number around the world once again rapidly increased.

During this phase the focus of industrial nations in Europe and the United States shifted to basic research, education, and the protection of plant species threatened due to global ecological problems.

The sustainable development of society and the search for harmonious cooperation between humans and nature has led to more and more comprehensive tasks for botanic gardens. It has become an undeniable fact that cities need more botanic gardens. Today there are more than 2,400 botanic gardens which together house approximately 80,000 plant species and over six million living plant specimens.

If one looks at the historical developmental of botanic gardens, the first gardens were herb gardens, or private, i.e. courtyard gardens. Prototypes of botanic gardens already existed over 2000 years ago in China, much earlier than in Europe. Several historic gar-

samt ca. 80.000 Pflanzenarten und über sechs Millionen lebendige Pflanzenpräparate beherbergen.

Wenn man die Entwicklungsgeschichte der Botanischen Gärten betrachtet, waren die ersten Gärten Kräutergärten, private Gärten bzw. Hofgärten. Prototypen botanischer Gärten existierten bereits vor über 2000 Jahren in China, viel früher als in Europa. Einige historische Gärten sind noch bis heute erhalten, wie zum Beispiel die Gärten des Chengde-Sommerpalastes, die des Yihe-Sommerpalastes im Norden und die Gärten von Suzhou im Süden. Sie gehören zur Elite der chinesischen Gärten, aber keiner von ihnen ist zu einem botanischen Garten auf Basis der modernen Botanik geworden. Dies ist kein Zufall, sondern spiegelt die Besonderheiten der Entwicklung der Wissenschaft und Technologie in China wider. Die modernen botanischen Gärten in China sind nicht aus der traditionellen Kultur abgeleitet worden, sondern resultieren aus westlichen Erkenntnissen. Es besteht keine geschichtliche Kontinuität zu den historischen Kräuter-, Blumen- bzw. Hofgärten. Selbstverständlich ist beiden Gartentypologien ihre jeweilige Bedeutung in der Gartenkunst gemein. Darüber hinaus aber gibt es keine Verbindungen zwischen der ursprünglichen chinesischen Gartentradition und den Botanischen Gärten der Gegenwart. Zu den größeren botanischen Gärten des 20. Jahrhunderts zählten der Zhongshan in Nanjing (1929), der Lushan (1934) und andere. Bis in die Entstehungszeit der Volksrepublik China ist nur der Botanische Garten Lushan größtenteils erhalten geblieben und hat die wirtschafts- und kriegsbedingten widrigen Umstände überdauert.

Nach Gründung der Volksrepublik China führte die Politik des wirtschaftlichen Aufbaus unter dem Leitsatz „wissenschaftlicher Inhalt, künstlerisches Aussehen" dazu, dass die Entwicklung der botanischen

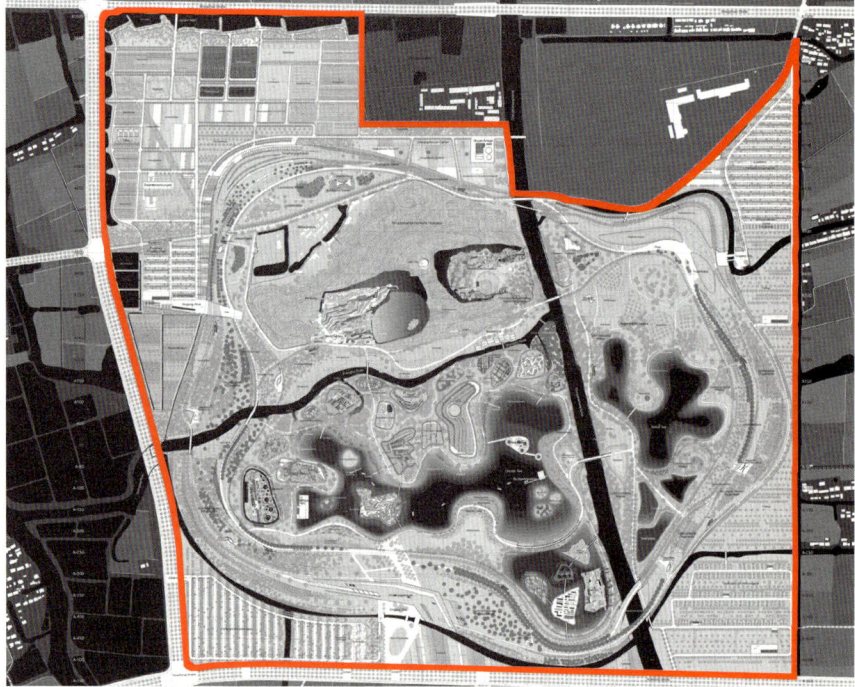

Gärten in China an Dynamik zunahm. Bis 1990 gab es bereits 102 botanische Gärten in China.

Zu Beginn des 21. Jahrhunderts haben immer mehr politisch Verantwortliche in Städten und Provinzen erkannt, dass botanische Gärten zum Schutz gefährdeter Pflanzenarten und deren Samen von großer Wichtigkeit sind, und dass diese Gärten eine große Bedeutung für die Entwicklung der Volkswirtschaft, der Urbanisierung und der Kultur haben. Botanische Gärten erhielten in China eine weitere Chance, sich zu entwickeln. Die Investitionen des Staates in Botanische Gärten haben bis heute stark zugenommen. Viele Städte bauen derzeit Botanische Gärten auf bzw. erweitern die bestehenden, darunter Shanghai, Ningbo, Taizhou, Wenzhou, Lin-an, Lijiang, Zhongdian, Shijiazhuang, Langfang, Xi-

Chenshan Botanischer Garten Shanghai, 200 ha The Chenshan Botanic Garden, Shanghai, 200 ha

Königlicher Botanischer Garten Kew, England, 120 ha Royal Botanic Gardens, Kew, England, 120 ha

dens have been preserved until today, for example the gardens of the Chengde Summer Palace and the Yihe Summer Palace in the north, and the gardens in Suzhou in the south. These belong to the elite of Chinese gardens, but none of these has become a botanic garden on the basis of modern botany. This is not a coincidence, and reflects the peculiarities of the development of science and technology in China. Modern botanic gardens in China have not been derived from traditional culture, but are instead the products of western knowledge. Ancient herb gardens and courtyard gardens have had no historic continuity, although both garden typologies have undoubtedly had an impact on garden design. In addition to this, however, there is no link between the original Chinese garden tradition and contemporary

botanic gardens. Among the largest botanic gardens of the 20th century were Zhongshan in Nanjing (1929), Lushan (1934), and others. At the time the People's Republic of China was founded, the only remaining botanic garden, which had to a great extent survived the long period of economic and war-related adversity, was in Lushan.

After the founding of the People's Republic of China the policy of economic expansion according to the guiding principle of "scientific content, artistic appearance" lent increased dynamism to the development of botanic gardens in the country. By 1990 there were already 102 such gardens in China.

At the beginning of the 21st century more and more of the officials in cities and provinces recognised that botanic gardens are useful for the protection of endangered plant species and that these gardens are also of great importance to the development of the economy, urbanisation, and culture. Botanic gardens in China were given another chance to develop. Cities have greatly increased their investment in botanic gardens in recent times. Many cities are currently either building new botanic gardens or increasing the size of existing ones, including those in Shanghai, Ningbo, Taizhou, Wenzhou, Linan, Lijiang, Zhongdian, Shijiazhuang, Langfang, Xiamen, Fuzhou, Quanzhou, Zhangzhou, Guangzhou, Dongguan, Nanning, Hefei, Jinan, Ordos and Taklamakan.

The area and extent of the botanic gardens in the above-mentioned cities are relatively large, usually between 50 and 60 hectares. Several are 100 to 200 hectares in size, and a few are even bigger. The volume of investment ranges from millions to billions of CNY. Developmental policy has been organised on the basis of the extended guiding principle of "scientific content, artistic appearance, cultural presenta-

amen, Fuzhou, Quanzhou, Zhangzhou, Guangzhou, Dongguan, Nanning, Hefei, Jinan, Ordos und Taklamakan. Fläche und Ausmaß der botanischen Gärten in den oben genannten Städten sind relativ groß, meist zwischen 50 und 60 Hektar, manche haben 100 bis 200 Hektar und einige sind sogar noch größer. Das Investitionsvolumen reicht von Millionen bis zu Milliarden CNY. Die Aufbaupolitik entwickelt sich anhand des erweiterten Leitsatzes „wissenschaftlicher Inhalt, künstlerisches Aussehen, kulturelle Vorführung" – eine Politik des wirtschaftlichen Wohlstandes. Heute gibt es nach weniger als 100 Jahren ca. 200 botanische Gärten in China, mit insgesamt über 17.000 Pflanzenarten, welche 50 bis 60 Prozent der ca. 30.000 Samenpflanzenarten Chinas entsprechen. Allerdings reichen diese 200 Gärten nicht aus, um Chinas reichhaltige und einzigartige Pflanzenressourcen zu schützen und zu erforschen. Chinas große Bevölkerung und die Urbanisierung, die immer mehr in den Vordergrund tritt, erfordern noch mehr botanische Gärten.

Shanghai ist eine der größten Städte Chinas und der Welt. Führende Politiker in Shanghai haben bereits vor langer Zeit realisiert, dass eine Großstadt von internationalem Rang nicht ohne botanische Gärten von hohem Niveau auskommt. Es ist absolut notwendig, in Shanghai ein hochwertiges Netzwerk botanischer Gärten aufzubauen, um dem Schutz und der Nutzung der Pflanzenressourcen und den Bedürfnissen der städtischen Entwicklung nachzukommen. Der gigantische Aufbau des Chenshan Botanischen Gartens in Shanghai spiegelt den Entwicklungstrend der botanischen Gärten auf der Welt und in China wider, er wird ein neues Kapitel in der chinesischen Entwicklungsgeschichte der botanischen Gärten eröffnen. Die Ereignisse in Shanghai haben bereits nationale und weltweite Aufmerksamkeit erregt. Um einen optimalen Plan zu finden, entstanden für den Chenshan Botanischen Garten acht nationale und internationale Entwürfe, die alle ein sehr hohes Niveau hatten. Aus der Sicht des Aufbaus und der ökologischen Sanierung stammt der beste Vorschlag von der deutschen Planungsgruppe Valentien. Angesichts weit verbreiteter ökologischer Schäden in der Umgebung soll der Botanische Garten nicht nur auf eine eigene Stabilität der Entwicklung, sondern auch auf die ökologische Sanierung in seiner Umgebung achten. Die Planungsgruppe Valentien berücksichtigt die tiefe Lage des Geländes und die schlechte Wasserqualität und schlägt einen „Grünen Ring" sowie Wasserreinigungssysteme vor, die ein solides Fundament für die nachhaltige Entwicklung des Botanischen Gartens mit seinen Teilgärten und Landschaftszitaten anbieten. Dies ist sehr lobenswert. Der gesamte Entwurf ist harmonisch, schön und außergewöhnlich.

Ich wünsche dem Aufbau des Chenshan Botanischen Gartens in Shanghai viel Erfolg!

tion" – a policy of economic prosperity. In less than 100 years approximately 200 botanic gardens have been built in China. They have over 17,000 species of plants, which represents 50 to 60 per cent of the circa 30,000 seed-bearing plants in China. These 200 gardens, however, are not sufficient for the protection and research of China's extensive and unique plant resources. The country's large population and its continuing urbanisation require even more botanical gardens.

Shanghai is one of the biggest cities in China and in the world. Leading politicians in Shanghai realised long ago that a world-class city of this size cannot get by without a first-rate botanic garden. It is absolutely vital that a network of high-quality botanic gardens be built in Shanghai in order to meet the requirements of plant protection and the use of plant resources. The gigantic scale of the Chenshan Botanic Garden in Shanghai, mirrors the level of development of botanic gardens around the world and in China and it opens a new chapter in the history of botanic garden development in China.

Events in Shanghai have already captured national and international attention. In order to arrive at an optimal plan, the Chenshan Botanic Garden Shanghai collected eight national and international designs, which were all of high quality. The best proposal, from the perspective of the layout and ecological redevelopment, came from the German Valentien planning group. In view of the widespread ecological damage in this area of the city, the botanic garden should not only pay attention to the stability of its own development, but to the ecological redevelopment of the entire surrounding area as well. The Valentien planning group considered the site's low-lying topography and its poor water quality and proposed a "green ring" as well as a water-purification system, which together build a solid foundation for the sustained development of the botanic garden, its theme gardens, and references to various landscapes. This is very commendable. The entire design is harmonious, beautiful, and exceptional.

I wish the development of the Chenshan Botanic Garden in Shanghai great success!

DIE WELTGESELLSCHAFT DER VEGETATION

THIES SCHRÖDER

Die Weltgesellschaft der Vegetation zu einem neuen Park zu formen, ist die Aufgabe des neuen Botanischen Gartens Shanghai in Chenshan. Ein künstlich aufgeschütteter Ring bildet die Basis für diese Vegetationsschau am Beispiel der Lorbeerwälder von Süd- und Nordamerika über Afrika und Europa bis hin nach Australien und Asien. In diesem Ring sind die Bauten des neuen Botanischen Gartens auf einzigartige Weise integriert. In dem vom Botanischen Ring umschlossenen Kernbereich finden sich Gärten zu einer Vielzahl an Themen und auch Attraktionen vom Labyrinth über die Medizin und die Naturgeschichte bis zur Osmanthuspflanze. Landschaft und Architektur finden zu einer neuen Symbiose – ein Zeichen nachhaltiger Entwicklung und eines neuen Verständnisses des Parks: dieser wird zum Ausdruck eines die Garten- wie die Baukunst integrierenden Ingenieurkunstwerkes.

Diese Konsequenz des europäischen Entwurfs hat die Stadt Shanghai und die überwiegend chinesisch besetzte Fachjury beeindruckt. Die künstlich entstehende neue Landschaft verbindet aktuelle Systemanforderungen einer ökologischen Entwicklung, beispielsweise die Qualität des Wassers in der Tieflandregion, mit den Erwartungen einer rasant wachsenden Stadtgesellschaft an ihre naturräumlichen Erholungspotenziale. Ein doppeltes Bildungsangebot entsteht mit dem neuen Botanischen Garten Shanghai: Die Landschaft als Ökosystem und als Gartenkunstwerk geht mit der Architektur und der Kulturgeschichte des Ortes eine Verbindung ein.

Das Wissen um die Pflanzen und ihre Gesellschaften weltweit wird in diesem 200 Hektar großen Botanischen Garten kultiviert. Die für China und die Region rund um Shanghai typischen, durch die rasante städtische Entwicklung bedrohten Pflanzenarten erhalten ein Refugium, ihre Erbinformationen werden bewahrt. Wissenschaftliche Forschung und das Erforschen der regional typischen Kulturlandschaft gehen eine gartenkünstlerisch gestaltete Symbiose ein: eine ästhetisch definierte, zugleich wissenschaftlich anregende Bildungslandschaft. Und ein Freizeitpark, der inmitten einer der weltweit größten und an Dynamik kaum zu übertreffenden Stadtregionen die urbanisierte Region mit der ursprünglichen Natur- und Kulturlandschaft verbindet.

Entworfen wurde der neue Botanische Garten Shanghai von den Landschaftsarchitekten Valentien + Valentien gemeinsam mit den Landschaftsarchitekten Straub + Thurmayr sowie den Architekten Auer + Weber + Assoziierte. Der Entwurf für den neuen Botanischen Garten Shanghai entstand auf der Basis eines globalen Diskurses. Die kulturellen Unterschiede zwischen europäischen und asiatischen Regionen, zwischen Gestaltungsvorstellungen, Haltungen zu Ökologie und Funktion im Entwurf sowie den Traditionen der Gartenkunst machen die Spannung aus, aus der die Ideen für den neuen Park gespeist werden. Eine regionale Identität? Sie soll in einem Park erkennbar werden, der Park wird eine Kulturlandschaft prägen. Doch diese muss nicht die Identität einer historischen Kulturlandschaft aufweisen, nicht museal und konservierend sein.

Auseinandersetzungen mit der Region Shanghai und dem konkreten Ort sowie Überlegungen zum Verhältnis dieses neuen Botanischen Gartens zur reichen Geschichte der chinesischen Gartenkunst hat es im Verlauf des Entwurfsprozesses viele gegeben: „Natürlich waren wir anfangs versucht – weitaus mehr als unsere Wettbewerbskonkurrenten aus China selbst übrigens –, die auf dem Gelände vorhandenen Siedlungselemente zu sichern und damit zu überhöhen. Auf uns wirkten diese einigermaßen exotisch, weshalb wir zunächst folkloristisch reagierten. Wir haben

THE GLOBAL PLANT COMMUNITY

THIES SCHRÖDER

Creating a park from the global vegetative community is the task of the new Shanghai Botanic Garden in Chenshan. An artificial elevated ring forms the basis of this vegetative show which includes examples of laurel forests from South and North America, Africa, Europe, Australia, and Asia. Within this ring the Botanic Garden's new buildings have been integrated in a unique way. In the core area surrounded by the botanic ring gardens with a variety of themes can be found, as well as attractions ranging from a labyrinth about medicine and natural history to Osmanthus plants. Landscape and architecture find a new symbiosis – a sign of sustainable development and a new understanding of parks: this has become an expression of an engineering masterpiece that integrates garden design and architecture.

This consistency of the European design impressed the City of Shanghai and the mainly Chinese jury. The artificially created new landscape links current system requirements of ecological development, for example the quality of the water in the lowland region, to the expectations of a rapidly growing urban society's recreational potential in natural areas. A twofold educational offer has been created with the new Botanic Garden Shanghai: The landscape as an ecosystem and as a piece of garden art is connected to the site through the architecture and cultural history of the area.

Knowledge of plants and their communities around the world is cultivated in this 200-hectare botanic garden. The area's plant species, which have been threatened due to the rapid urban development so typical of China and the region around Shanghai, are thus given refuge, and their genetic information is preserved. Scientific research and an investigation of the cultural landscape typical to the region enter into a designed garden-architectural symbiosis: an aesthetically defined, and at the same time academically stimulating, educational landscape. And a park, in the midst of one of the world's largest and most dynamic urban regions, that connects the developing region with its original natural and cultural landscape.

The new Shanghai Botanic Garden was designed by landscape architects Valentien + Valentien together with landscape architects Straub + Thurmayr, as well as the architects Auer + Weber + Assoziierte.

The design for the new Shanghai Botanic Garden originated on the basis of a global dialogue. The cultural differences between European and Asian regions, between ideas about design, attitudes towards ecology and function in design, as well as traditions of garden design create a tension which nourishes ideas for new parks. Regional identity? This should be apparent in a park as it will characterise a cultural landscape. And yet, this park shouldn't exhibit the identity of a historic cultural landscape, should not be a museum nor a conservationist gallery.

During the design process there were many discussions and debates concerning the Shanghai region and the concrete location as well as deliberations about the relationship of this new botanic garden to the rich history of Chinese garden design. "At first, of course, we were tempted – much more than our competitors from China, incidentally – to preserve elements of the existing settlement on site and to accentuate them. They appeared fairly exotic to us, which is why we initially reacted in a folkloric manner. But during the course of the competition we quickly learned that the special task of a botanic garden, a park about the world's ecosystems, requires a completely different solution: The park must have a design approach that reflects the location of this botanic garden and the particular tradition of Chinese garden design it represents. This should not, how-

dann aber im Verlauf des Wettbewerbs schnell gelernt, dass die ganz besondere Aufgabe eines Botanischen Gartens, eines Parks der Ökosysteme der Welt also, eine ganz andere Antwort erfordert: Der Park muss eine gestalterische Haltung haben, die reflektiert, wo dieser Botanische Garten liegt und in welcher Tradition der chinesischen Gartenkunst er steht. Doch diese ist nicht als Anleitung zur Bewahrung, sondern als Herausforderung zur Erneuerung zu verstehen. Wir haben auf diesem Weg unendlich viel gelernt." (Christoph Valentien)

Ort und Region Der neue Botanische Garten Shanghai liegt inmitten der Erweiterungszone einer der weltweit am schnellsten wachsenden Regionen. Eine bauliche Entwicklung unvorstellbarer Geschwindigkeit ereignet sich als Ausdruck der Modernisierung eines Staates wie einer Weltzone.

„Kopf der neun Gipfel in den Wolken" wird eine Gruppe von Hügeln genannt, die über die weitgehend flache Topografie des Verwaltungsgebietes Shanghai im Mündungsgebiet des Jangtse hinausragen. Das Areal des neuen Botanischen Gartens ist wie 90 Prozent der Fläche Shanghais durch ländliche und vorstädtische Siedlungsmuster geprägt, während rund zehn Prozent des 6340 Quadratkilometer großen Verwaltungsgebietes zur eigentlichen Kernstadt gehören.

Das Wissen um die Geschichte und die Sozialstruktur eines Ortes ist für den Entwurf ebenso notwendig wie die intensive Auseinandersetzung mit dem Ort selbst, seinem Klima, seiner Identität und seiner Zukunft. Bauen innerhalb der Vorgaben eines Gartens? Der Maßstab für die bisherige Entwicklung Shanghais ist diese Haltung gerade nicht. Vielmehr wurde schlicht abgerissen, planiert, geschüttet und errichtet. In wenigen Jahren und Jahrzehnten entwickelte sich und

entwickelt sich weiterhin eine in Bauten gefasste Entwicklungsdynamik, wie sie in Europa kaum mehr vorstellbar erscheint.

Seine wirtschaftliche Prosperität verdankt Shanghai einem von den Autoritäten Chinas strikt geplanten und verordneten Kapitalismus. Shanghai ist heute kapitalistische Vorzeigemetropole, eine Stadt mit 18 Millionen Einwohnern, in der die langwelligen Entwicklungen anderer Städte der Welt im Zeitraffer nachvollzogen und überholt werden. Der Weg Chinas sieht solche Konzentrationsräume wirtschaftlicher Dynamik vor, Shanghai markiert als bedeutendste Industriestadt der Volksrepublik China die wichtigste dieser Entwicklungszonen.

So sind auch die Flächen rund um den neuen Botanischen Garten für eine rasche weitere Bebauung vorbereitet. Das von zahlreichen Flüssen und Kanälen durchzogene Gebiet wurde in den letzten Jahren durch ein Raster breiter Straßenzüge erschlossen. Shanghai hat die Kapazitäten seines Straßennetzes seit den 1990er-Jahren fast verdoppelt.

Zugleich sind in Shanghai die Grenzen der Wachstumsdynamik inzwischen erkennbar. Umweltbelastungen nehmen stark zu, die Verkehrstrassen sind chronisch überlastet. Die Stadtentwicklung reagiert planvoll: Stadtökologie und Umweltschutz bekommen in kürzester Zeit eine wichtige Bedeutung zugesprochen. Der Entwurf für den neuen Botanischen Garten kam hier zur richtigen Zeit. Der Park ist in seiner Idee der Integration der Ingenieurstechnik, vor allem der biologischen Wasserreinigung, in eine im internationalen Maßstab innovative Architektur- und Gartenarchitektursprache und in einen Ort der Bildung wie der Unterhaltung weltweit vorbildlich.

Die Themen auf diesem Weg der Entwicklung der Metropolregionen sind die global gleichen. Blicken wir in den Newsletter einer internationalen Fachzeitschrift

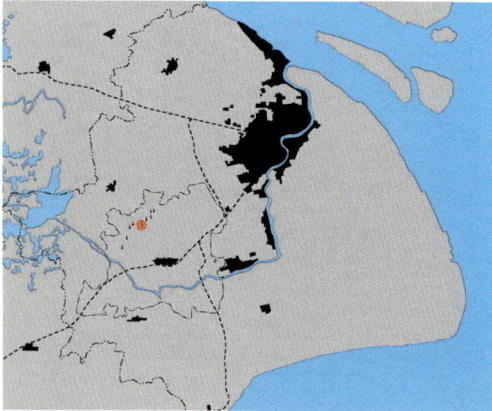

Lageplan Chenshan Botanischer Garten Shanghai Ground plan, Chenshan Botanic Garden, Shanghai

ever, be understood as a rationale for preservation, but instead as a challenge for renewal. By looking at it this way we learned a great deal." (Christoph Valentien)

Site and Region The new Shanghai Botanic Garden is located in the midst of an expanding zone in one of the world's fastest growing regions. Structural development of unimaginable speed is occurring as an expression of the modernisation of a country, of an entire part of the world.

"Head of the Nine Peaks in the Clouds" is the name of a group of hills which tower above the predominantly flat topography of Shanghai, which is located in the delta area of the Yangtze River. Up to now this area has had a rural structure, similar to 90 per cent of the Shanghai area, which is characterised by rural and suburban development patterns. Ten per cent of Shanghai's 6340 square kilometres belong to the actual core urban area.

Knowledge of the history and social structure of an area is an essential part of design, just as is an intensive examination of the area itself, of its climate, its identity and its future.

Building within the constraints of a garden? The scale of the development in Shanghai up to now has not exactly adopted this approach. On the contrary, simple demolition, levelling, mounding and erection have taken place. In just a few years and decades a developmental dynamism visible in the form of new construction has taken place, and is continuing on a scale that is hardly imaginable in Europe anymore.

Shanghai owes its economic prosperity to authoritarian China's brand of capitalism, which has been strictly planned and put into action. Today Shanghai is a showcase capitalist metropolis, a city of 18 million inhabitants which has imitated and rapidly overtaken other developing cities around the world. China's way forward calls for this type of concentrated economic zone, and Shanghai, as the main industrial city in the People's Republic, is the most important of these developmental zones.

Thus the properties surrounding the new botanic garden have also been prepared for rapid construction. The area, which is crossed by various rivers and canals, has been opened up and given an additional grid of broad streets in the last few years. Shanghai has almost doubled the capacity of its road networks since the 1990s.

In the meantime, however, the limits of this dynamic growth in Shanghai have been recognized. Environmental pollution has greatly increased and the roads are constantly overloaded. City planning has reacted in a well-planned manner: Urban ecology and environmental protection have quickly become more important. The design for the new botanic garden was carried out at exactly the right time. The main idea of the park is to integrate engineering technology, especially biological water purification, with innovative and international architecture and landscape architecture and with a centre of education and entertainment. The results should be exemplary worldwide.

The themes involved with the development of metropolitan regions are the same around the world. Just take a look at a newsletter from an international journal of landscape architecture produced in China: At the turn of the year 2007-2008 the leading article is about a competition for students, addressing the subject of "The design and planning of small

24

zur Landschaftsarchitektur, die in China produziert wird: Ende 2007 steht an erster Stelle ein Wettbewerb für Studierende, die zum Thema „The design and planning of small size sustainable human habitat projects" arbeiten sollen. "Great changes have taken place on urban and rural living environment, the broken and shabby living blocks, crowded work environment and the continual weak natural environment. The International Association for Humane Habitat (IAHH) aims at promoting and advancing healthy and harmonious human habitat by means of student design competition. ... The contestants should design some public facilities, which can supply 200–500 job opportunities. Its shape and work site depend on its local, social, economical, cultural and environmental conditions." Diese Wettbewerbsauslobung hätte gleich lautend ebenso auf einen Wettbewerb zum demografischen Wandel in Deutschland oder zu einem Siedlungsprojekt in den Niederlanden gepasst, aber auch zu einem innovativen Denkprozess in den USA nach der Immobilienkrise. Nachhaltigkeit wird global als verbindende Frage an Arbeit und Wirtschaft, Ökologie und Umwelt, Bildung und Kultur definiert. Der wegweisende Park der Gegenwart hat sich in diesem Feld zwischen regionaler Tradition und globaler Zukunft zu positionieren.

Die Olympischen Spiele 2008 in China finden in dem zitierten Newsletter ebenso ihre Erwähnung wie die veränderte chinesische Position zum Klimawandel: "Wen (Premier Wen Jiabao) emphasized that the Chinese Government put more emphasis on climate change. China has actively taken measures to save energy on the basis of its own sustainable strategy, and has now made great progress. China will have an active and responsible attitude within the Bali Convention Conference and will make its contributions to the success of the conference."

Ereignisse wie die Olympischen Spiele oder die EXPO Shanghai 2010 sind heute wesentliche Motoren weltweiter Entwicklung und Erneuerung. Das Ereignis hat den Plan als Handlungsorientierung abgelöst. Ereignis wie Plan aber benötigen ein Programm, eine Haltung. Für die EXPO Shanghai wird dieses Programm auch durch den Botanischen Garten verkörpert: „Wir schaffen mit dem Botanischen Garten einen nachhaltigen Park und ein wegweisendes Gartenkunstwerk, das sich an den Regelwerken der Ökologie orientiert und gerade deshalb eine neue künstliche Landschaft unter Integration der Architektur schafft. Unsere Landschaft als ein Abbild der Welt sollte einen Beitrag leisten zum Wissen über das Ökosystem Welt und zum nachhaltigen Umgang mit diesem." (Christoph Valentien)

Die Gestaltungsideen für den neuen Botanischen Garten bieten Orientierung für die chinesische Anwendung der globalen Diskussion um Chancen und Grenzen des Wachstums. Was gut durchdacht und geplant ist, kann Gutes beitragen zur Zukunft unserer Entwicklung. Daher bezieht der Botanische Garten Position – die Position der Landschaftsarchitekten und Architekten. Eine Haltung, die weit mehr umfasst als das Abenteuer, einen Park für Shanghai entwickeln zu dürfen, weit mehr als das Abenteuer China, das zurzeit so sehr in aller Munde ist.

Die Planungsgruppe hat einen Beitrag zur globalen Landschaftsarchitektur verfasst und in den Wettbewerb gegeben. Anfangs passierte, was häufig passiert im Abenteuer Wettbewerb wie im Abenteuer China. Eine Jury tagt, eine Reihenfolge wird festgelegt, Ideen werden prämiert oder angekauft und manche Idee wird auch übernommen.

Beim Wettbewerb für den neuen Botanischen Garten Shanghai wurde das mit dem dritten Preis gewürdigte Entwurfsteam mit der weiteren Entwurfsplanung be-

size sustainable human habitat projects." "Great changes have taken place on urban and rural living environment, the broken and shabby living blocks, crowded work environment and the continual weak natural environment. The International Association for Humane Habitat (IAHH) aims at promoting and advancing healthy and harmonious human habitat by means of student design competition. ... The contestants should design public facilities, which can supply 200—500 job opportunities. Their shape and work site depend on their local, social, economical, cultural and environmental conditions." This type of announcement could easily have been for a competition about demographic change in Germany or for a residential estate in the Netherlands, or even for initiating an innovative thinking process in the United States in the wake of the real estate crisis. Sustainability is now defined worldwide as something which connects work and the economy, ecology and the environment, and education and culture. Contemporary leading edge parks have to position themselves in this domain between regional tradition and a global future.

The 2008 Olympic Games are also mentioned in the newsletter, as is the new Chinese position concerning climate change: "Wen (Premier Wen Jiabao) emphasized that the Chinese Government put more emphasis on climate change. China has actively taken measures to save energy on the basis of its own sustainable strategy, and has now made great progress. China will have an active and responsible attitude within the Bali Convention Conference and will make its contributions to the success of the conference."

Events like the Olympic Games or the 2010 Expo in Shanghai are now essential motors of worldwide development and renewal. Events have replaced plans as a way of doing business. But both events and plans require a programme, an approach. For the Shanghai Expo the Botanic Garden has become the embodiment of this programme: "With the Botanic Garden we are creating a sustainable park and a groundbreaking example of garden design that is orientated towards a policy of ecology, and is thus a new artificial landscape with integrative architecture. Our landscape as a reflection of the world should contribute to knowledge about the world as an ecosystem, and to a sustainable way of dealing with it." (Christoph Valentien)

The ideas behind the design of the new botanic garden serve as orientation concerning the Chinese view of the global discussion about the chances and limits of growth. That which is well thought-out and planned can contribute to the future of our development. Hence, the botanic garden has taken up a position, i.e., the position maintained by landscape architects and architects. This is an approach which involves much more than just the adventure of being able to develop a park for Shanghai, much more than the adventure of China itself, which everyone is talking about.

The planning group developed a contribution to global landscape architecture and entered the competition. In the beginning, things occurred which are just as common to the adventure of a competition as they are to the adventure of China. A jury meets, a sequence is decided upon, ideas are given prizes or honourable mentions, and some ideas are taken up and put into practice.

In the competition for the Shanghai Botanic Garden the design team that won third prize was commissioned to continue designing the park, provided they work together with a partner firm from Shanghai. And thus the chance arose to turn a distinctive concept

26

auftragt, unter Einbeziehung eines Partnerinstitutes aus Shanghai. So entstand die Chance, ein profiliertes Konzept umzusetzen, das internationalen Vergleichen standhält. Es ist eben nicht allein ein neuer Erholungspark in der Metropolregion Shanghai, nicht allein ein neuer Botanischer Garten, sondern auch ein EXPO-Projekt, das hier entsteht. Die weltweite Beachtung ist diesem Garten sicher, der umzusetzende Entwurf sollte, so die Intention der Jury, diesem globalen Publikum gewachsen sein. (Übrigens war die Jury dieses Wettbewerbs nicht von internationalen Fachkollegen dominiert, sondern die Entscheidung fiel in Shanghai durch die Vertreter der dortigen Freiraumplanung und Gartenkultur.)

Dass die Planungsgruppe Valentien für den Botanischen Garten nun über den Wettbewerbsentwurf hinaus eine vollständige und weitgehend verbindliche Entwurfsplanung entwickeln konnte, ist der Qualität des vorgelegten Entwurfs geschuldet, aber ebenso Ausdruck der Wertschätzung für Prof. Christoph Valentien in China. Über Jahre hinweg entstand durch Vorträge, Wettbewerbsbeiträge und Hochschulkooperationen, oft gemeinsam mit Dietmar Straub, eine Basis für die künftige Zusammenarbeit. Wichtiger und unverzichtbarer Partner in diesem Zusammenhang war Yiju Ding, der im wörtlichen wie im übertragenen Sinne Übersetzer war zwischen den beiden Welten. Die Möglichkeit, den Chenshan Botanischen Garten über den Wettbewerb hinaus weiter gestalten zu können, basiert auf der langjährigen Vorarbeit und einem gewachsenen gegenseitigen Vertrauen. Man hatte auf diesen Reisen ebenso wie anhand der Projekte des Büros in Deutschland deutlich machen können, dass die Planungsgruppe nicht per se für die beste Lösung steht, sondern dass im Team und in allen Entwurfsphasen stets um die bessere Lösung gerungen wird,

getragen von einer gereiften eigenen Haltung und einem Respekt für einen Ort, eine Kultur, ihre Traditionen und ihre Visionen. Auf diesem Weg entsteht nun mit dem neuen Botanischen Garten Shanghai ein Meilenstein in der Welt der Gartenkunst.

Gegenseitige Beeinflussung Gegenüber dem planerischen Engagement in China mischten sich in der Vergangenheit in Europa Vorbehalte und Faszination. Der neue Botanische Garten ist Ausdruck einer neuen Phase: Es ist inzwischen allen Seiten offenbar ernst mit der internationalen Zusammenarbeit. Die noch nicht ganz beendete Phase der Zitate, hier wie dort übrigens, wenn ich an den chinesischen „Garten des wiedergewonnenen Mondes" als Herzstück der „Gärten der Welt" in Berlin-Marzahn denke oder aber an Projekte wie die britische Stadt Thamestown oder die deutsche Stadt in China – diese Phase also der kulturellen Globalneugierde, die auch etwas Disneyhaftes hat, wird nun offenbar abgelöst durch eine Phase der Kooperation und gegenseitigen Beeinflussung.

Das kann tatsächlich noch sehr spannend werden. Wie kann nun eine solche gegenseitige Beeinflussung gelingen? Da ist zum einen die Vorgeschichte. So wie einst alles Persönliche politisch war, ist heute alles Globale auch lokal. Man muss sich kennenlernen, vor Ort, ob nun im Gartenbauamt von Shanghai oder an der Uni in München. Die persönliche Ebene der Kontakte ist die einzige Chance, den Gedankentransfer, der ja immer auch ein beidseitiger Kulturschock ist, kreativ umsetzen zu können. Projekte wie dieses benötigen Vermittler, im besten Sinne.

Da ist aber auch die Sicherheit, trotz der notwendigen kulturellen Vermittlung und des Respektes vor der chinesischen Gartentradition die eigene Idee konsequent zu verfolgen und weiterzuentwickeln.

into reality, to create a project that would hold up to international comparison. For what is being created is not only a new public park in the Shanghai metropolitan region, and not only a new botanic garden, but is an Expo project as well. There is certain to be great global interest in this park and the jury wanted to be sure the realisation of this design would live up to all expectations. (Interestingly, the jury was not dominated by a panel of international professionals, and decisions in Shanghai were made by representatives of the local landscape planning profession and gardening associations.)

The fact that the Valentien planning group was able to develop complete and largely binding design development documents for the botanic garden from the competition's schematic design is due to the quality of the initial design, and is also an expression of the esteem with which professor Valentien is held in China. A basis for this cooperation was created through the lectures, competition entries and university collaboration, often together with Dietmar Straub, that had occurred over the years. An important and absolutely essential partner in these cooperative efforts was Yiju Ding, who, in both a literal and a figurative sense, served as a translator between both worlds. The possibility to continue the design of the botanic garden beyond the competition was a result of many years of preliminary work as well as deep-rooted mutual trust. Through these trips and based on the firm's projects in Germany it became apparent that the planning group didn't merely stand for the best solution per se, but that within the team and in all phases of the design an active debate would take place in order to find the best solution, based on a mature approach and respect for a site, a culture, its

traditions, and its visions. In this way the new Shanghai Botanic Garden was created, a milestone in the world of garden design.

Mutual Influence In the past Europe has been both reserved and fascinated by the spate of planning activities in China. The new botanic garden is an expression of a new phase: The seriousness of international cooperation has now become apparent to all sides. The phase of replication, present here as well as there and not yet completely at an end (the Chinese "Garden of the Regained Moon" for instance, which is the centrepiece of the "Gardens of the World" in Berlin-Marzahn, or projects like the British city of Thamestown, or the German City in China), this phase of global curiosity, which has something of Disneyland about it, is apparently being replaced by a phase of cooperation and mutual influence.

This could actually become quite exciting. How can this kind of mutual influence succeed? For one thing there is past history. Just as everything personal was once political, today everything global is also local. People must get to know each other locally, whether it's at the Department of Public Parks in Shanghai or at the University in Munich. The personal level of contact provides the only chance to have a creative exchange of thoughts, which is always a culture shock for both sides. Projects like this require a mediator, in the best sense of the word.

On the other hand, however, there is the certainty, despite the necessary cultural mediation and respect for Chinese garden traditions, of being able to consistently pursue and develop one's own ideas further.

ENTWICKLUNGSSTRATEGIE DES BOTANISCHEN GARTENS IN SHANGHAI

HU YONGHONG

Mit dem stetig zunehmenden Tempo der modernen Wissenschaft und Technologie wird die Funktion des Botanischen Gartens von Tag zu Tag umfassender. Daher ist es fortwährend Shanghais Wunsch, einen weltweit fortschrittlichen, modernen Botanischen Garten aufzubauen, der Shanghai als eine internationale Großstadt ausweist.

Aufbau und Entwicklung des Botanischen Gartens in Shanghai haben eine lange Geschichte. Bereits in den 30er-Jahren des 20. Jahrhunderts waren im Inneren des Yaofeng-Parks in Shanghai (der heutige Zhongshan-Park) ein kleiner Botanischer Garten und ein Zoo eingerichtet worden, die später jedoch an Bedeutung verloren. In den 50er-Jahren sah die Regierung von Shanghai östlich und westlich des She-Berges in Songjiang ca. 400 Hektar Land für den Aufbau eines Botanischen Gartens vor, später wurde das Projekt aus wirtschaftlichen Gründen aufgegeben. Mitte der 70er-Jahre schlugen ältere Mitarbeiter der Longhua-Baumschule vor, auf Basis der Baumschule einen Botanischen Garten aufzubauen. Dieser Vorschlag wurde angenommen. Mithilfe der Universität für Forstwirtschaft Beijing wurde der Vorläufer des heutigen Botanischen Gartens in Shanghai aufgebaut.

Im Jahr 2003 beantragte die Bezirksregierung von Songjiang bei der Stadtregierung, im Gebiet des She-Berges einen neuen Botanischen Garten in Shanghai aufbauen zu dürfen. Das Projekt wurde 2004 von der Stadtregierung genehmigt. Ende März 2007 begann der Aufbau. Das 200 Hektar Land umfassende Projekt wird vor der Eröffnung der Expo 2010 vollendet und der Öffentlichkeit zugänglich gemacht werden.

Der schon existierende Botanische Garten in Shanghai liegt im südwestlichen Teil der Stadt. Es handelt sich um eine Parkanlage, die wissenschaftliche For-

schung, Populärwissenschaft und Touristik vereint. Sein Vorgänger ist die Longhua-Baumschule, deren Umwandlung zum Botanischen Garten 1974 begann. Der Botanische Garten hat eine Fläche von 81,86 Hektar. Sein Ausstellungsbereich setzt sich zusammen aus den Bereichen der Pflanzlichen Evolution, dem Bereich der Künstlichen Ökologie, dem Bereich des Umweltschutzes, dem Demonstrationsbereich der Grünen Zone und dem Touristikbereich der Huangmu-Ahnenhalle sowie fünfzehn Themengärten verschiedener pflanzlicher Kategorien. Dabei nimmt der Bereich der Evolution der Pflanzen den größten Stellenwert ein.

Der Botanische Garten in Shanghai war von Beginn an bestrebt, verschiedene neue Pflanzenarten einzuführen, die spezifisch an das Klima und den hiesigen Boden angepasst sind. Bis jetzt hat der Botanische Garten über 5000 Pflanzenarten gesammelt (einschließlich Mutationen und einiger Sorten des Gartenbaus), wovon mehr als 120 seltene Arten vom Aussterben bedroht sind. Er bemüht sich, Hängepflanzen, Pflanzen, die besonders geeignet im Kampf gegen Umweltverschmutzung sind, Zierblumen, Blattpflanzen und einheimische Bäume zu verbreiten,

Alter Botanischer Garten Shanghai, Tee-Haus Old Botanic Garden Shanghai, teahouse

THE DEVELOPMENT STRATEGY FOR THE BOTANIC GARDEN IN SHANGHAI

HU YONGHONG

The function of the botanic garden has become progressively more comprehensive as a result of the steadily increasing tempo of modern science and technology. And thus it has long been Shanghai's wish to build an outstanding, advanced, and modern botanic garden which would promote the city as an international metropolis.

The development of the Botanic Garden in Shanghai has a long history. A small botanic garden and zoo have existed in Yaofeng Park (currently known as Zhongshan Park) since the 1930s, but are no longer of much importance. During the 1950s the government in Shanghai set aside approximately 400 hectares of land for the construction of a botanic garden east and west of the She Mountain in Songjiang, but the project was later abandoned for economic reasons. In the mid-1970s elder employees at the Longhua nursery suggested creating a botanic garden on the grounds of the nursery. This suggestion was accepted. With the help of the University for Forestry in Beijing the precursor to today's Botanic Garden in Shanghai was built.

In 2003 the Songjiang district government applied to the municipal government for permission to build a new Botanic Garden in Shanghai near the She Mountain. The project was approved by the municipal government in 2004. Construction began in 2007. The 200-hectare project will be completed before the opening of the Expo 2010 and made accessible to the public.

The existing botanic garden in Shanghai is located in a southwest section of the city. It consists of a park which combines scientific research, popular science, and tourism. Its predecessor is the Longhua Nursery, which was converted into a botanic garden in 1974. It's 81.86 hectares include various exhibition areas (showing the evolution of plants, artificial ecol-

ogy and environmental protection), a demonstration area for the "green zone" and an area for tourists at the Huangmu ancestors' hall, as well as 15 gardens for different categories of plants. The most important of these is the area demonstrating the evolution of plants.

From the very beginning the botanic garden in Shanghai endeavoured to introduce new species of plants which were specifically adapted to the climate and local soils. The botanic garden has collected over 5000 species of plants (including mutations and various horticultural examples), including more than 120 rare species that are threatened with extinction. It is also involved in breeding hanging plants, plants which are especially suitable in the fight against environmental pollution, ornamental flowers, foliage plants, and native plants for the greening of urban roofs, walls, balconies, and windowsills.

The botanic garden in Shanghai organizes a range of activities every year, including, for example, the Spring Flower Show. Permanent exhibitions with a variety of themes linked to the seasons attract many tourists. The botanic garden encourages visitors to take part in practical activities and offers training in popular science issues. In this way it serves as a green classroom for students from primary and secondary schools.

As a result of accelerating urbanisation in Shanghai the quantity of different plant species is being continuously reduced. The establishment of the new Botanic Garden in Chenshan is of extraordinary importance for the protection of wild plants as a resource and as a potential for research. The construction of the new botanic garden is an important contribution to the greening of the city; it embodies the general developmental level of the economy, culture, science, and the middle classes. As the gross domestic

die für die Begrünung der städtischen Dächer, Wände, Balkone und Fensterbänke benötigt werden.

Jedes Jahr wird im Botanischen Garten in Shanghai eine Vielzahl an Aktivitäten, wie zum Beispiel die Frühlingsblumenausstellung, durchgeführt. Dauerausstellungen, die Jahreszeiten mit verschiedenen Themen verbinden, ziehen Touristen an. Der Botanische Garten animiert Besucher zur Teilnahme an praktischen Tätigkeiten und bietet ein vielfältiges populärwissenschaftliches Training an; er wird so zu einem grünen Klassenzimmer für Schüler aus Grundschulen und Gymnasien.

Infolge der beschleunigten Urbanisierung von Shanghai sinkt die Anzahl der Pflanzenarten beständig. Die Einrichtung des neuen Botanischen Gartens in Chenshan hat eine außerordentliche Bedeutung für den Schutz der Wildpflanzen als Ressource und als Forschungspotenzial. Der Aufbau dieses botanischen Gartens ist ein wichtiger Beitrag zur Begrünung der Stadt, er verkörpert das allgemeine Entwicklungsniveau der Wirtschaft, Kultur, Wissenschaft und des Bürgertums. Mit dem Anstieg des Pro-Kopf-Bruttoinlandsprodukts von Shanghai und Umgebung auf über 7000 US-Dollar stieg auch das Bedürfnis der Bevölkerung an populärwissenschaftlicher Bildung. Der Chenshan Botanische Garten folgt schlussendlich dem Wettbewerbsentwurf der Planungsgruppe Valentien aus Deutschland. Er berücksichtigt nicht nur die Besonderheiten des Veranstaltungsortes, sondern auch die Prinzipien der Zhuanshu-Kalligrafie. Der äußere Rahmen des chinesischen Zeichens für Garten symbolisiert einen grünen Ring, die drei Radikale innerhalb des Rahmens symbolisieren jeweils Berge, Wasser und Pflanzen im Park, sodass die traditionelle chinesische Gartenkunst abgebildet wird.

Der Aufbau des Chenshan Botanischen Gartens wird die Entwicklung des jetzigen Gartens nicht schwächen, sondern ergänzen. Seine Hauptorientierung ist „der Schutz und die Erforschung von regionalen strategischen Ressourcen und von seltenen und bedrohten Pflanzen". Xue Dayuan sagte: „Manchmal wird das Schicksal eines Landes und einer Nation durch das Entdecken einer einzelnen Wildpflanzenart geändert." Der bestehende Botanische Garten in Shanghai nutzt seine langjährige Erfahrung, um sich auf die Erforschung der feinen Gartenkunst und die Züchtung von neuen Pflanzen zu spezialisieren. Die Blumenschau in Shanghai dient als Mittel zur Befriedigung des Verlangens der Bevölkerung nach Blumen und Schönheit.

Das Ziel der beiden Botanischen Gärten ist ein hohes internationales Niveau und Ansehen; dies braucht selbstverständlich Zeit und Mühe.

Alter Botanischer Garten Shanghai, Bonsai-Garten Old Botanic Garden Shanghai, Bonsai Garden

product in Shanghai and its neighbouring areas has grown to over 7000 US dollars per person, the need to educate the population about popular science has increased as well. In the end the Chenshan Botanic Garden in Shanghai chose the design of the Valentien planning group from Germany. This concept not only takes into account the distinctive features of the venue, but also the principles of Zhuanshu calligraphy.

The outer frame of the Chinese symbol for garden symbolises a green ring, the three radicals within the frame represent mountains, water, and plants in a park, thus depicting the design principles of traditional Chinese gardens.

The construction of the Chenshan Botanic Garden will not weaken the development of the current garden, but instead will complement it. Its main orientation is the "protection and researching of strategic regional resources and of rare and threatened plants." Xue Dayuan once said: "Sometimes the fate of a country and a nation is changed through the discovery of a single wild plant species." The existing Botanic Garden in Shanghai will use its long-standing experience to specialise in research about the beautiful and subtle sides of garden design and the breeding of new plants. The flower show in Shanghai will serve as a means to satisfy the population's desire for flowers and beauty.

The goal of both botanic gardens is to win respect and to reach a high international level; this will naturally require both time and effort.

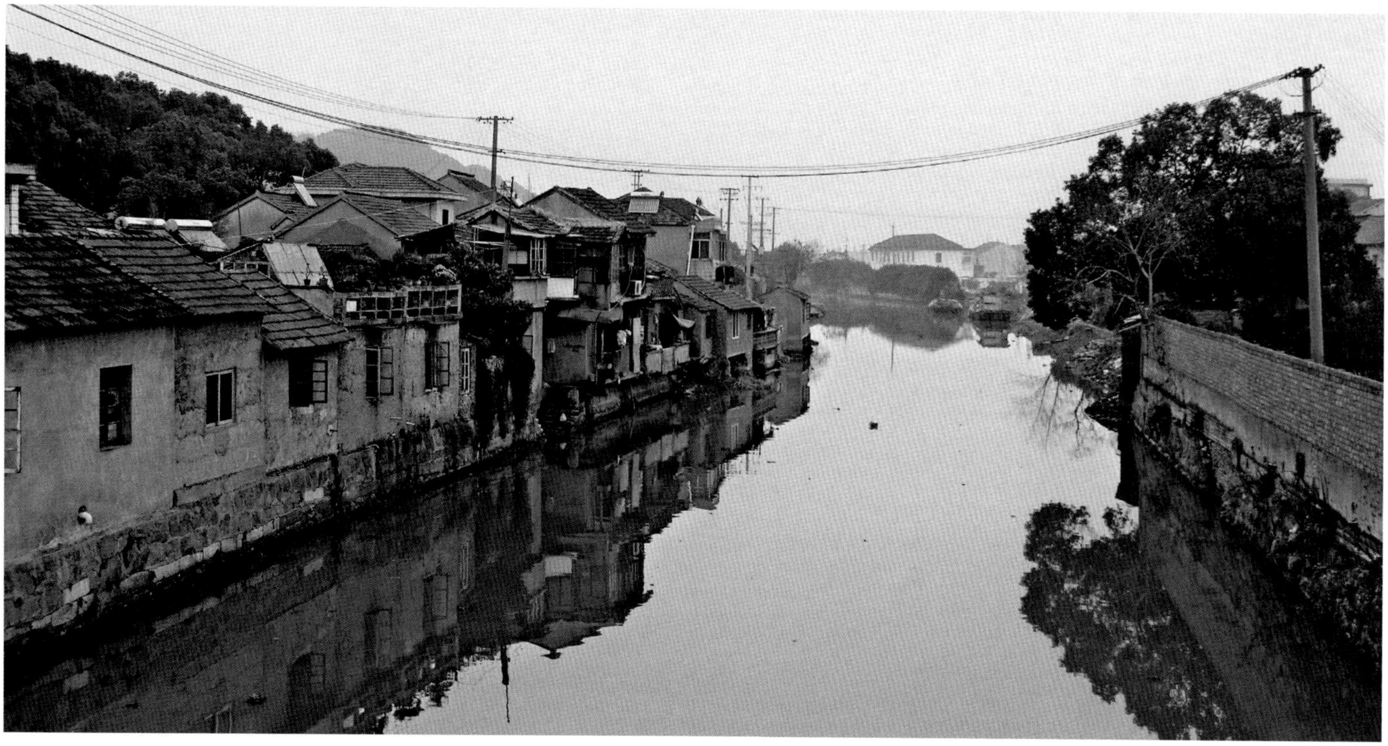

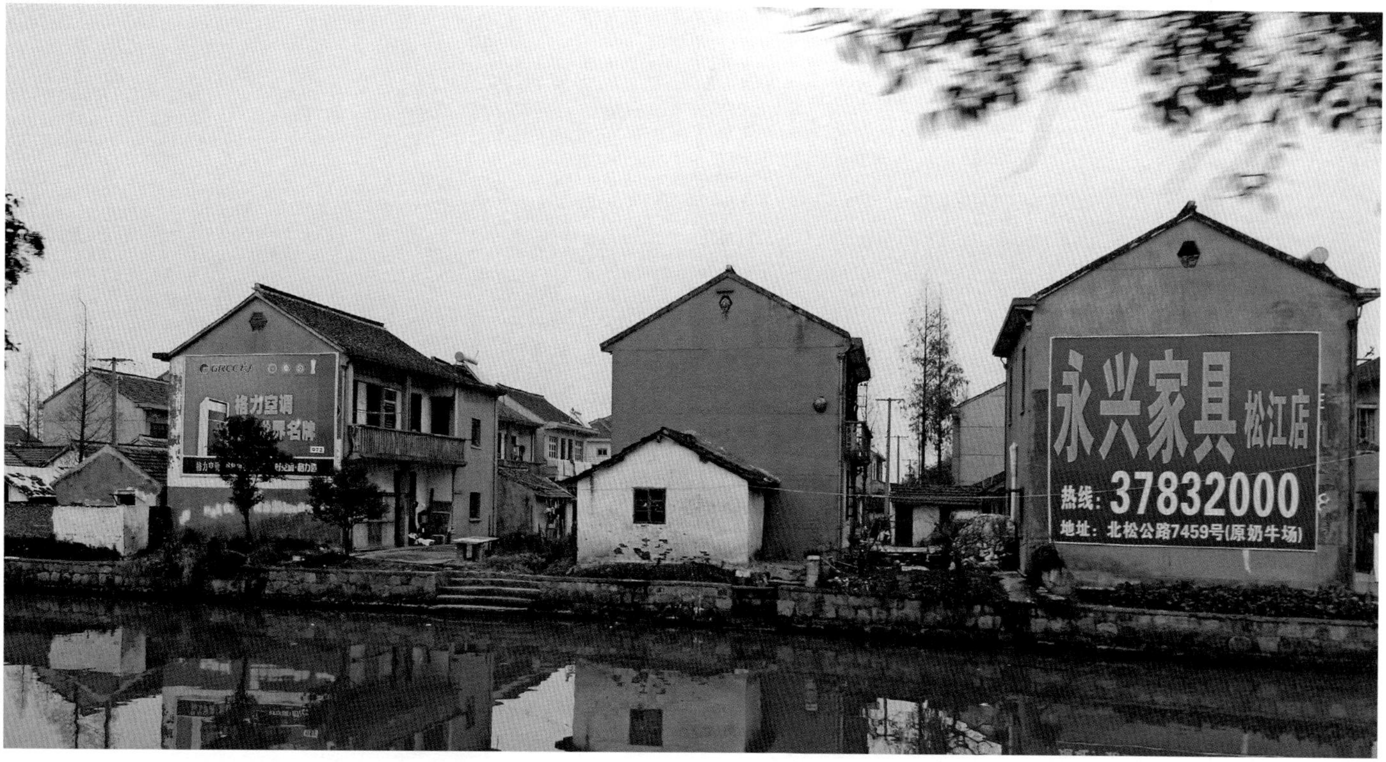
格力空调
永兴家具 松江店
热线: 37832000
地址: 北松公路7459号(原奶牛场)

CHENSHAN BOTANISCHER GARTEN SHANGHAI

DONATA UND CHRISTOPH VALENTIEN

„Die erste Nachricht von Chinas Gärten gibt uns der Venezianer Marco Polo in dem Berichte von seiner Reise, die er als Kaufmann in den Jahren 1272—93 unternahm. Marco Polo gelangte damals an den Hof Kublai Khans, des großen Mongolenkaisers. Er sah den schönen Tierpark [...], zu dem man nur, wie in allen chinesischen Gartenanlagen, durch den Palast gelangte [...]. Er sah den Palast des Großkhans in Cambalu, dem heutigen Peking, und schilderte seine doppelte Umwallung: [...] durch die üppige Vegetation führen gepflasterte und erhobene Fußsteige, von denen der Regen abfließt, so daß sie nie schmutzig sind und die Vegetation ringsumher immer bewässert ist. An der Ecke der Umwallung [...] liegt ein schöner Fischteich, der von einem Fluß durchflossen wird, bronzene Gitter am Ab- und Zufluß hindern die Fische am Fortschwimmen. In dem eigentlichen Palastgarten bewundert er besonders einen volle hundert Schritt hohen, künstlichen Hügel, der an der Basis ungefähr eine Meile [...] beträgt; dieser ist aus der Erde, die man aus dem See gegraben hat, errichtet und mit schönsten immergrünen Bäumen besetzt; denn sobald seine Majestät erfährt, daß an irgendeinem Platze ein schöner Baum wächst, läßt er ihn mit allen Wurzeln und der umgebenden Erde ausgraben, wie groß und schwer er auch ist, und durch Elefanten auf diesen Hügel schaffen, so daß er die schönste Sammlung von Bäumen in der Welt hat."[1]

Botanische Gärten – Orte für Pflanzen In Europa begann die „Jagd nach dem grünen Gold" mit den großen Entdeckungsreisen im 16. Jahrhundert. „Tollkühne und auch Besessene fuhren aus, um nach seltenen und unbekannten Pflanzen zu suchen, die in der Heimat häufig mit Gold aufgewogen wurden. Es bedurfte eines gewissen Wagemuts, wenn es galt, den reichen Gartenfreunden oder anspruchsvollen Handelsgärtnern Orchideen aus den brasilianischen oder fernöstlichen Urwäldern zu holen." (Dietmar Straub)

Zur gleichen Zeit begannen die Universitäten, botanische Gärten für Forschung und Lehre anzulegen. Einer der ersten und bekanntesten ist der botanisch-pharmazeutische Garten von Padua. Im Kern ist der kleine Garten noch heute so erhalten, wie er 1545 entstand: ein verwunschener stiller Ort inmitten der Stadt. In einem streng geometrischen Grundriss wird eine bescheidene Kollektion von Pflanzen ausgestellt.

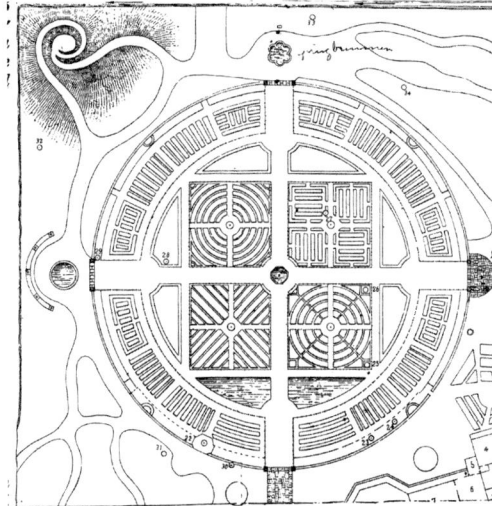

Stich Grundriss Botanischer Garten Padua An engraving of the layout of the Botanic Garden of Padua

Dies war der Beginn der modernen Botanik. Im Botanischen Garten in Padua wurden die Pflanzen zum Gegenstand des Systematisierens des Weltwissens.

Vor allem das 18. Jahrhundert war in Europa dann von großer Bedeutung für die Entwicklung von Gärten und botanischen Sammlungen. Carl von Linné entwickelte mit seiner binären Nomenklatur (1756) die Voraussetzung für eine systematische Sammlung und Demonstration von Pflanzen.

1 aus from: Marie Luise Gothein, Geschichte der Gartenkunst, Zweiter Band, Von der Renaissance in Frankreich bis zur Gegenwart, Herausgegeben mit Unterstützung der Königlichen Akademie des Bauwesens in Berlin, Verlegt bei Eugen Diederichs in Jena, 1914, S. 319, 322
The Book of Marco Polo, ed. by H. Yule I, p. 326ff.

THE CHENSHAN BOTANIC GARDEN IN SHANGHAI

DONATA AND CHRISTOPH VALENTIEN

"The first ample accounts of the gardens of China are given by Marco Polo the Venetian, in the history of his travels undertaken in the years 1272—93. When Marco Polo arrived at the court of Kublai Khan, the great Mongol emperor, he saw the deer-park [...] which one could only reach [...]. He also saw the palace of the Great Khan at Cambalu (Kambalu), and described its double row of encircling walls [...]. The footpaths were paved and somewhat raised, so that the rain ran off them; thus they were never dirty, and the vegetation through which they ran was always well watered. At the corner of the circumvallation [...], there was a fish-pond. A river ran through it, with gratings at each end to prevent the fish swimming through. In the garden proper Marco Polo especially admired an artificial mound, fully a hundred feet high, standing on a base of about a mile [...]. This mound was made of earth dug out for the lake, and on it there were handsome evergreen trees, As soon as the emperor heard of a beautiful tree anywhere, he had it dug up with all its roots and a great deal of earth, and conveyed to this mound by elephants, however heavy it might be: thus he acquired the finest collection of trees in the world."[1]

Botanic Gardens – Places for Plants In Europe the "hunt for the green gold" began with the long journeys of discovery in the 16th century. "The reckless and obsessed set out to search for rare and unknown plants, which were often worth their weight in gold back at home. A certain amount of daring was required when it came to getting orchids from Brazilian and Far East jungles for wealthy gardeners or demanding commercial gardeners." (Dietmar Straub) At the same time universities began to develop botanic gardens for research and teaching. One of the first and most famous is the botanic-pharmaceutical garden in Padua. The core of the small garden, built in 1545, has been preserved in its original form: a quiet enchanted place in the centre of the city. A humble collection of plants is exhibited in a strict geometric layout.

This was the beginning of modern-day botany. In the Botanic Garden of Padua the plants became part of the classification of world knowledge.

In Europe the 18th century was especially important in the development of gardens and botanic collections. With his binary nomenclature (1756) Carl von Linné developed the prerequisite for a systematic collection and demonstration of plants.

Alexander von Humboldt opened up the world of plants through his long research expeditions. Many exotic plants were imported to Europe, and through his systematic work it became possible to assign them in a geographic fashion. Large arboretums were established, not only in botanic gardens but also in many parks. The Royal Botanic Gardens Kew in London were the first to be detached from a university and made accessible as a public park.

Botanic gardens have been among the large design commissions that landscape architects dream of ever since the Botanic Garden in Padua shaped the world's view of botany and thus our understanding of the variety of plant genera and species. And ever since Paxton's glass buildings in London's Kew Gardens we know that this is also the case for architects. It is the utopian idea of being able to give the world, concentrated in one space, a scientific and aesthetic look all at the same time.

The new Shanghai Botanic Garden in Chenshan documents another time, a new understanding and experiencing of the world. We have become enthusiastic travellers, know the different parts of the world from our own experience, or at least from documen-

38

Alexander von Humboldt erschloss mit seinen großen Forschungsreisen die Welt der Pflanzen neu.
Viele Exoten wurden nach Europa importiert, durch
die systematische Arbeit Humboldts waren sie nun
auch geografisch zuzuordnen. Es entstanden große
Arboreten, nicht nur in Botanischen Gärten, sondern
auch in vielen Parks. Dabei waren die Königlichen
Botanischen Gärten Kew in London die ersten, die
von der Hochschule losgelöst auch als öffentlicher
Park zugänglich gemacht wurden.

Seit der Botanische Garten in Padua das Weltbild
der Botanik und damit unser Verständnis von der
Vielfalt von Gattungen und Arten der Pflanzen prägt,
gehört ein Botanischer Garten zu jenen großen Ent-
wurfsaufgaben, von denen der Landschaftsarchitekt
träumt. Und seit Paxtons Glashäusern in Londons
Kew Gardens wissen wir, dass dieses ebenso für Architekten gilt. Es ist der utopische Gedanke, der Welt
– auf einen Raum konzentriert – zugleich ein wissen-
schaftliches wie ein ästhetisches Gesicht geben zu
können.

Der neue Botanische Garten Shanghais in Chenshan
dokumentiert eine andere Zeit, ein neues Verstehen
und Erleben der Welt. Wir sind begeisterte Reisende
geworden, kennen die Erdteile aus eigenem Erleben
oder zumindest aus Filmdokumenten und gehen mit
anderen Augen in einen solchen Garten. Das Wissen über die Pflanzen der Erde ist gewachsen und
wächst ständig weiter. Zugleich aber wachsen auch
die Erkenntnis von der globalen Gefährdung der
Ökosysteme und das Wissen um den dramatischen
Verlust an Pflanzenarten. Der Botanische Garten
Shanghai kann als Garten der Neuzeit nicht mehr
nur wissenschaftliche Pflanzensammlung sein, sondern er muss die Vielfalt pflanzlicher Lebensgemeinschaften beispielhaft abbilden. Er ist ein Ort, an dem
man gewissermaßen in der Welt spazieren geht.

Damit steht der Garten in der Tradition chinesischer
Gartenkunst, in der nicht die einzelne Pflanze im Mittelpunkt steht, sondern der Garten Abbild der Landschaft ist, des Weltganzen, symbolisch verdichtet
in Berg und Tal, Wasser und Fels, Wüste und Wald,
Schatten und Licht.

An diese Tradition knüpfen wir an und entwerfen dennoch neue Bilder. Denn es soll Verständnis geweckt
werden für die Ökosysteme der Erde in ihrer Schönheit und in ihrer Gefährdung. Der Forderung, verant-
wortlich umzugehen mit der Natur, muss zunächst im
Garten selbst nachgegangen werden, sie muss sich
dem erholungsuchenden Besucher des Botanischen
Gartens vermitteln, lehrreich, unterhaltsam und un-
vergesslich.

Die Kultur des Unterschieds Mit abendländi-
schen Augen haben wir uns dem Ort genähert, den
Garten vor dem Hintergrund unserer eigenen Garten-
träume gesehen. Zugleich waren wir neugierig, woll-
ten begreifen, was anders ist, welche Träume und
Hoffnungen sich in China mit Landschaft, mit Garten
verknüpfen.

Der europäische Garten, althochdeutsch *garto*, meint
das Umzäunte, planmäßig Angelegte. Die Grenze,
der Zaun, die Hecke, die Mauer sind erste gesicherte
Merkmale eines Gartens. Sie schützen vor einer als
bedrohlich empfundenen Natur und geben Gebor-
genheit. Gartenkultur setzt das Innen und Außen in
Beziehung und behandelt so auch immer das Verhältnis zwischen dem Vertrauten und dem Fremden.
„[J]eder Garten hegt die Liebe zum Anderen, zum
Nichteigenen: ‚Und eben darum enthält der Garten
gerade nicht diejenige Blume, die vor den Gittern frei
vorkommt, sondern die einer Landschaft des Andersseins, einer Sehnsucht. Das Gitter trennt zwei
Welten, wozu wäre es sonst da? [...] Gartenblumen,

taries, and see such a garden with different eyes. Knowledge of the world's plants has increased and constantly continues to do so. At the same time, however, an understanding of the global danger to ecosystems and knowledge about the dramatic loss of plant species is growing. The Botanic Garden in Chenshan can, as a modern-day garden, no longer just function as a scientific plant collection, but must exhibit the great variety of biotic communities. It is a site where one goes for a walk, as it were, in the world at large.

And thus the garden remains within the tradition of Chinese garden design, where the plant itself is not the centre of attention, but the garden is a reproduction of the landscape, the entire world, symbolically reduced to mountains and valleys, water and rocks, desert and forest, shadows and light.

We continue with this tradition, but at the same time create new images. The garden should awaken an understanding for the earth's ecosystems in their beauty and in the danger they are exposed to. The demand to deal with nature in a responsible way must first be dealt with in the garden itself; it has to communicate, instruct, entertain and be memorable to visitors who are seeking relaxation.

The Culture of Difference We approached the site with western eyes; saw the garden from the background of our own garden dreams. At the same time, we were curious, and wanted to understand what was different, which dreams and hopes are linked to the landscape, to the garden, in China.

The European garden, *garto* in Old High German, is something that is fenced in, and laid out according to a plan. The boundary, the fence, the hedge, and the walls are the first verifiable signs of a garden. They protect against a nature perceived as threatening,

lending a sense of security. Garden culture creates a correlation between inside and outside and always approaches the relationship between the familiar and the foreign in this way.

"Every garden nurtures the love of that which is different, of that which does not belong: And that's the reason why the garden doesn't include the very same flower that grows freely on the other side of the enclosure, but instead a landscape of otherness, of longing. The fence separates two worlds; why else would it be there? […] Garden flowers, as one defensively calls them, are wild flowers in other countries."[2]

The Chinese garden is also defined by these elements of otherness, but as opposed to the European garden, it doesn't exclude the landscape, on the contrary, it accepts it in a stylised form, embraces the earth and water, mountain and sky.

Garden design in China, as opposed to Europe, is characterized by great continuity. The architectural garden that dominated European garden design for centuries was only a rare and briefly assumed style. The goal is not always to find new images of gardens, but instead to refine the image, the elegance of the organisation, the wealth of variations concerning the use of elements and symbols. Chinese garden design is the representation of landscapes, of ideal landscapes; in the end it is a form of landscape painting. For thousands of years, independent of the individual situation or the size of the garden, similar elements have been brought into play: mountain and valley, lakes, rivers, and forests. Architectural symbols are used in much the same way the: the bridge, the teahouse, the pavilion are orchestrated as a concealed resting place or as a *point de vue* for a particular section.

Mountains and rocks are particularly revered. Mar-

2 Gustav Seibt, 2003. Seibt quotes Rudolph Borchard's writing about the "passionate gardener", written after 1937 and published in 1967 by Marie Luise Borchardt. "A book about the humanisation of nature, which, in the tracing of discoveries and trade routes plants from all parts of Europe secretly travelled along, discusses the principle of a mixture of the apotheosis", "a cultural theory which celebrates nothing more than the limitless creolisation, the eternal recreation of culture from conquered nature".
Gustav Seibt, "Das brechende Herz des besseren Mannes. Rudolf Borchardt während des Dritten Reiches", in Merkur. Deutsche Zeitschrift für europäisches Denken, vol. 57, no. 6 (650) (June 2003):465-479.

wie man sie verteidigend nennt, sind die wilden Blumen anderer Länder'."[2]

Auch der chinesische Garten ist durch diese Elemente des Andersseins definiert, aber im Unterschied zu den europäischen Gärten schließt er Landschaft nicht aus, im Gegenteil, er nimmt sie in stilisierter Form in sich auf, umschließt Erde und Wasser, Berg und Himmel.

Anders als in Europa ist die Gartenkunst Chinas von großer Kontinuität geprägt. Der architektonische Garten, der über Jahrhunderte die europäische Gartenkunst beherrschte, war ein nur selten und flüchtig übernommenes Zitat. Nicht die Erfindung immer neuer Gartenbilder ist das Ziel, sondern die Verfeinerung in der Darstellung, die Eleganz der Zuordnung, das variantenreiche Spiel mit Elementen und Symbolen. Chinesische Gartenkunst ist die Darstellung von Landschaften, von Ideallandschaften, sie ist letztlich eine Spielart der Landschaftsmalerei. Seit Jahrtausenden werden ähnlich und fast unabhängig von der jeweiligen Situation oder der Größe des Gartens die immer gleichen Elemente zitiert: Berg und Tal, Seen, Flüsse und Wälder und gleichermaßen wiederkehrend die baulichen Symbole, die Brücke, das Teehaus, der Pavillon, sei es als verborgener Aufenthaltsort oder als *point de vue* einen bestimmten Ausschnitt inszenierend.

Berge und Felsen werden ganz besonders verehrt. Martini spricht in seinem chinesischen Atlas von einem seltsamen Aberglauben der Chinesen die Berge betreffend: „Sie erforschen die Psychologie eines Berges, seine Formation, seine Adern, wie sonst Astrologen den Himmel."[3]

Die Verwendung von Pflanzen in wiederkehrender Symbolik gleicht eher formelhaften Bildern: die Allee aus Trauerweiden, zum See führend; die windgeformte Kieferngruppe auf der Bergklippe; der Lotussee mit gebogenen Brücken, die Gruppe von Päonien am Rande der Terrasse.

Wir haben diese Bilder und Elemente aufgenommen und versucht, sie zu übersetzen in ein Konzept, das den Erfordernissen eines modernen Botanischen Gartens gerecht wird und eine formale Sprache spricht, die wir selbst glaubwürdig vertreten können.

Der Ort Shanghai ist auf Schwemmland gebaut, liegt in einem gewaltigen Delta, das der Jangtse an seiner Mündung in das Ostchinesische Meer gebildet hat.

In weithin ebener Landschaft verweisen neun Granithügel von etwa 100 Metern Höhe auf den älteren Untergrund der Landschaft; sie waren vermutlich Inselberge vor der Küste, ehe der Jangtse das Meer hat verlanden lassen. Das Land liegt nur knapp über dem Meeresspiegel, das Grundwasser steht hoch an, Hochwasserschutz ist ein zentrales Thema.

Das Klima ist subtropisch maritim, feuchtkalt im Winter, gelegentlich gibt es Frost. Der Sommer dagegen ist schwülwarm mit Temperaturen von 28–30 Grad Celsius und einer Luftfeuchtigkeit, die nicht selten bis zu 100 Prozent beträgt. Die Niederschläge liegen bei 1120 Millimetern im Jahr, 50 Prozent der Menge regnet es allein in den Sommermonaten, wenn auch Taifune die Gegend heimsuchen.

Noch ist feuchtes Schwemmland vorherrschend, „das Wasserland", von zahlreichen Kanälen und Entwässerungsgräben durchzogen. Landwirtschaft, Gemüseanbau, Obst und intensive Fischzucht prägen das Bild, dazwischen finden sich dörfliche Streusiedlungen. Aber die Landschaft wandelt sich in dramatischer Geschwindigkeit. Die Satellitenstadt Songjiang im Südwesten Shanghais wird ein neuer Siedlungsschwerpunkt werden; allenthalben wachsen Großsiedlungen und Villenquartiere aus dem Bo-

2 Gustav Seibt 2003. Seibt zitiert Rudolf Borchardts Schrift über den ‚Leidenschaftlichen Gärtner', entstanden ab 1937 und 1967 herausgegeben von Marie Luise Borchardt: „Ein Buch über die Vermenschlichung von Natur, das in der Nachzeichnung der Entdeckungen und Handelswege, welche Pflanzen aus allen Erdteilen in Europa heimisch gemacht haben, dem Prinzip Mischung die Apotheose schreibt", „eine Kulturtheorie, die nichts anderes feierte als die schrankenlose Kreolisierung, die ewige Neuschöpfung der Kultur aus überwundener Natur." Gustav Seibt: „Das brechende Herz des besseren Mannes. Rudolf Borchardt während des Dritten Reiches". In: Merkur. Deutsche Zeitschrift für europäisches Denken. Jg. 57, Heft 6 (650) vom Juni 2003 S.465-479, hier 475)

3 aus: Marie Luise Gothein, Geschichte der Gartenkunst, Zweiter Band, Von der Renaissance in Frankreich bis zur Gegenwart, Herausgegeben mit Unterstützung der Königlichen Akademie des Bauwesens in Berlin, Verlegt bei Eugen Diederichs in Jena, 1914, S. 330; Martinus Martini, Novus Atlas Sinensis, 1655, in Blaeu, Weltatlas (Geographie Blaviane), Amsterdam 1667, S. 17

tini speaks in his Chinese atlas of a strange super-stition the Chinese have regarding mountains: "They research the psychology of a mountain, its formation, and its veins, similar to what astrologists do with the sky."[3]

The use of plants in repeating symbolism somewhat resembles stereotyped images: an avenue of weep-ing willows that leads to a lake, a wind-blown group of pines on a mountain ledge, a lotus-filled lake with an arched bridge, or a group of peonies on the edge of a terrace.

We accepted these images and tried to translate them into a concept that does justice to the require-ments of a modern botanic garden and uses a formal language that we can credibly advocate using.

The Site Shanghai is built on alluvial land, located in an immense delta formed by the Yangtze River as it flows into the East China Sea.

In a largely flat landscape nine granite hills rise to about 100 metres in height, referring back to the landscape's older subsurface; they were presum-ably inselbergs just off the coast before the Yangtze turned this part of the sea into land. The land lies just slightly above sea level and the groundwater is close to the surface, making flood management a major issue.

The climate is classed as subtropical maritime, and is cold and damp in the winter, with an occasional bout of frost. The summer, by contrast, is warm and humid with temperatures of 28—30 degrees Celsius and a relative humidity which often rises to 100 per cent. Total annual rainfall amounts to 1,120 millimetres; 50 per cent of this falls during the summer months, when typhoons also strike the area.

A moist alluvial landscape still prevails, a "water landscape" which is crossed by numerous canals and drainage channels. Agriculture, the cultivation of vegetables and fruit, and intensive fish farming char-acterise the image, and in between widely scattered villages can be found. The landscape, however, is changing at a dramatic pace. Songjiang, a satellite city in south-western Shanghai, will become a centre of new residential development. Large new housing estates and exclusive residential areas are springing up everywhere along with a new university centre and business parks and, quickly built prior to all this, wide-spreading, generously proportioned grids of streets that cut through the plain in large squares. The land in this area has obviously been prepared for further construction and the Botanic Garden will soon be an internal park in this continuously growing metropolis.

The nine hills, "Head of the Nine Peaks in the Clouds", are witnesses to the history of the earth and land-scape. They will survive as a distinctive sign, as a holy place. But soon they will no longer dominate the open plain, and instead will be inlays of nature in this large city's sea of buildings.

The Design A strong formative intervention was necessary in order to create a coherent location in such heterogeneous surroundings. We decided to view the garden as a solitary element, quite accentu-ated, and made no attempt to adapt it to its envi-rons.

The spatial composition is simple. A large sculptur-ally graded ring surrounds an inner garden and one of the nine hills. The ring is a symbol for the world, and within it the mountain, the water, and the sky re-flected in it are the defining symbols and spatial mo-tifs. The powerful dynamism of the spatial elements gives an orientation in the geographical as well as the aesthetic sense.

3 Marie Luise Gothein, Geschichte der Gartenkunst, Zweiter Band, Von der Ren-aissance in Frankreich bis zur Gegenwart (Eugen Diederichs, Jena, 1914): 330. Martinus Martini, "Novus Atlas Sinensis", in Geographie Blaviane (Amsterdam, 1667): 17.

den, ein neues Universitätszentrum, Gewerbezonen und vorauseilend weitmaschige, großzügige Straßenraster, die quadratförmig die Ebene durchschneiden. Die Landschaft ist sichtbar Bauerwartungsland und der Botanische Garten wird schon bald ein innen liegender Park sein in der noch immer wachsenden Metropole.

Die neun Hügel, „Kopf der neun Gipfel in den Wolken", sind Zeugenberge der Erd- und Landschaftsgeschichte. Sie werden überdauern, als Merkzeichen, als heilige Orte. Aber sie werden bald nicht mehr die offene Ebene dominieren, sondern Intarsien von Natur im Häusermeer der Großstadt sein.

Der Entwurf Es war eine starke gestalterische Intervention nötig, um in einer so heterogenen Umgebung einen lesbaren Ort zu schaffen. Wir entschieden uns dafür, den Garten als Solitär zu sehen, herausgehoben, ohne den Versuch der Anpassung.

Die räumliche Komposition ist einfach. Ein großer, skulptural modellierter Ring umschließt den inneren Garten und einen der neun Hügel. Der Ring steht als Symbol für den Erdenkreis, im Inneren sind der Berg, das Wasser und der sich darin spiegelnde Himmel die bestimmenden Symbole und Raummotive. Die kraftvolle Dynamik der Raumelemente gibt Orientierung im geografischen wie im ästhetischen Sinn.

Dabei war die Zeichenhaftigkeit des Entwurfes nicht intellektuelles Spiel, sondern Ergebnis der Vertiefung in die Aufgabe und in den Ort. Zugleich gibt diese räumliche Komposition das Konzept vor, das diesen Garten von anderen zeitgenössischen botanischen Gärten unterscheidet. Auf dem hohen Ring mit seinen Hängen und Verebnungen führt eine pflanzengeografische Wanderung durch die Lorbeerwälder aller Kontinente. Das Herz des Gartens bildet die große Seenlandschaft, in der Wasserpflanzen naturnah und

künstlerisch dargestellt werden. Etwa 35 inselförmig aus dem feuchten Grund gehobene Themengärten zeigen schließlich die züchterische Vielfalt blühender und grüner Pflanzen.

Alle wichtigen Bauten, die Empfangsgebäude mit Ausstellungshallen, die Glashäuser und das Forschungszentrum für Botanik, sind in die große Skulptur des Ringes integriert und bilden mit ihm eine morphologische Einheit. Die Verbindung natürlicher und gebauter Strukturen ist jedoch weitergehend. Hinter der ästhetischen Annäherung stand das Ziel, symbolisch und real die friedliche Koexistenz von Natur und Architektur darzustellen und Nachhaltigkeit und Kreislaufwirtschaft auf allen Ebenen zu erproben.

Man könnte meinen, ein Garten sei per se Symbol und Beispiel gebauter Nachhaltigkeit. Ein botanischer Garten ist dies keineswegs. Pflanzen aus aller Herren Länder auf einen Standort zu konzentrieren, heißt, mit erheblichem Aufwand künstlich die Standortbedingungen dieser Länder herzustellen und mit ständiger Pflege das Überleben der kostbaren Fremden zu gewährleisten.

Im Chenshan Botanischen Garten heißt dies, in einem von Natur aus feuchten, sumpfigen Gelände trockene Standorte zu schaffen oder tonigen Untergrund, wo nötig, in sauren Moorboden zu verwandeln und die verschmutzten Gewässer so zu reinigen, dass auch empfindliche Wasserpflanzen in den künstlich geschaffenen Seen gedeihen. Und wie in allen botanischen Gärten sind Häuser aus Glas mit Heizung, Klimaanlage, künstlichem Wassermanagement und hohem Energieaufwand erforderlich. Dennoch war es das Ziel, ein nachhaltiges System aufzubauen, von der Gebäudeplanung bis hin zu Bau und Betrieb des Gartens.

Der Wall und die aus dem feuchten Untergrund herausgehobenen Themengärten sind zwar nur mit er-

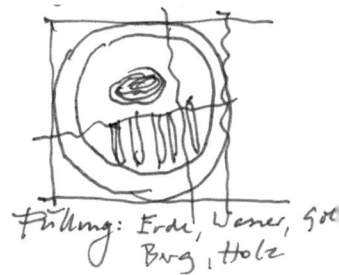

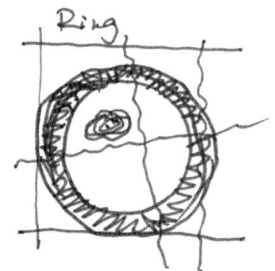

Ideenskizze
An initial sketch

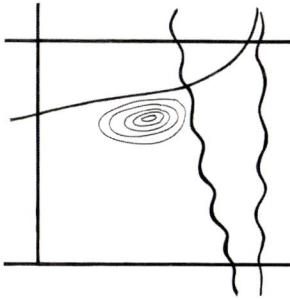

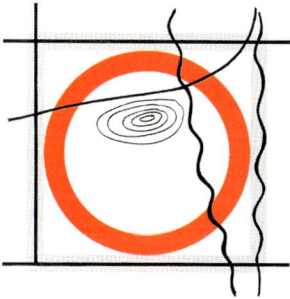

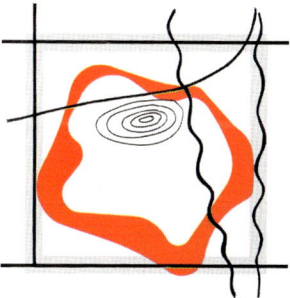

The symbolism of the design, however, was not so much an intellectual matter as it was the result of immersing ourselves in the task at hand and the site itself. At the same time, the spatial composition determines the concept that distinguishes this garden from other contemporary botanic gardens. On the elevated ring, with its slopes and level areas, a vegetative-geographical pathway leads through laurel woods from different continents. The heart of the garden is the large lakeland area, where aquatic plants are exhibited in both natural and artistic settings. And finally, approximately 35 island-shaped theme gardens arising out of the wetlands demonstrate the breeding diversity of flowering and green plants.

The most important buildings, i.e. the reception building with its exhibition halls, the greenhouses, and the botanical research centre, are all integrated into the large sculptured ring and together with this form a morphological unit. But the relationship between natural and built structures does not end here. The goal behind the aesthetic approach was to illustrate the peaceful coexistence of nature and architecture in a symbolic and realistic way, and to test sustainability and circular economics at all levels.

One might be tempted to think that a garden is per se a symbol and example of built sustainability. This by no means pertains to a botanic garden. To concentrate plants from all four corners of the earth in one location means going to great expense to artificially create the local conditions of those countries and to provide these valuable exotica with constant maintenance to assure their survival.

In the Chenshan Botanic Garden this means creating arid conditions in a humid, marshy area or transforming clayey subsoil, where needed, into acidic peaty soil, and purifying the polluted local water to a degree that sensitive aquatic plants can thrive in the artificially constructed lakes. And, as in all botanic gardens, there are buildings made of glass with heating and air conditioning systems, artificial water management that will use great amounts of energy. And yet, the goal here was to develop a sustainable system, from the planning of the buildings to the operation of the garden.

Building the embankment and the theme gardens which rise up out of the marshy subsoil occurred at great cost, and required a great deal of energy, but at least the soil excavated from the large lakes could be used on site. In this way expensive road haulage to the surrounding region was avoided, which will remain the goal in the future. Biomass and waste materials will be collected and converted into energy for cooling and heating in an on-site biogas plant. In addition to the biogas plant there are also demonstration plots for energy plants. The heavily polluted local water will be purified in a constructed wetland area and fed into the lakes. Rainwater will be collected and used for everyday operations. The garden will thus be able to function as an example and demonstration project for sustainable architectural development far beyond the actual site.

Topos and Topography The graded relief, the hills and embankments that merge with the buildings, the lake, and the stylised landscape – clearly the fabulous Olympic Park in Munich (architects Behnisch & Partner with Günther Grzimek as landscape architect) is a spiritual point of reference for this composition. The difference between the two projects, however, in the basic approach and the formal manifestation, is equally as important. Grzimek interlaced the park and the city, drew the large roads into the grading concept, and used viewing axes as connections to the city.

heblichem Aufwand, auch an Energie, zu realisieren, immerhin aber konnte dadurch das Erdmaterial aus den großen Seen vor Ort direkt wieder eingebracht werden. Aufwendige Massentransporte ins Umland wurden so verringert. Auch im künftigen Betrieb wird dies angestrebt. In einer eigenen Biogasanlage sollen Biomasse und Abfälle gesammelt und in Energie für Kühlung und Heizung umgewandelt werden. Im Anschluss an die Biogasanlage liegen die Demonstrationsfelder für Energiepflanzen. Das stark verschmutzte Wasser der Umgebung wird in einer Schilfkläranlage gereinigt und den Seen zugeführt, Niederschlagwasser wird gesammelt und für den Betrieb verwendet. Der Garten kann damit als Beispiel und Demonstrationsvorhaben nachhaltiger baulicher Entwicklung über den Standort hinaus wirken.

Topos und Topografie Das modellierte Relief, die Hügel und Wälle, die mit den Gebäuden verschmelzen, der See, die stilisierten Landschaften – natürlich ist der großartige Olympiapark in München der Architekten Behnisch & Partner mit Günther Grzimek als Landschaftsarchitekt ein geistiger Bezugspunkt dieser Komposition. Ebenso wesentlich aber ist der Unterschied, im Ansatz wie im formalen Ausdruck. Grzimek verwob Park und Stadt, bezog die großen Straßen in die Modellierung ein, suchte mit Blickachsen die Verbindung zur Stadt.

In Chenshan gibt es die große Stadt noch nicht. Die Blickpunkte sind nicht definiert, die ästhetischen Qualitäten unbekannt. Sicher ist nur: Die Landschaft, auf die man heute vom Ring aus schaut, wird verschwinden. So ist statt der Verschmelzung mit der Stadt die Ausgrenzung gewählt worden, das archaische Bild des Gartens, der *hortus conclusus*. Erst wenn der Wall durchquert ist, öffnet sich der Blick über den See in die Weite des paradiesischen Gartens.

Dort beginnt ein Spiel mit Widersprüchlichem, mit dem Verhältnis von Wildnis und Garten. In der Modellierung des Geländes wird die Künstlichkeit der Szenerien betont. Die Linien des Walles, seine Neigungen, seine Brechungen sind unübersehbar artifiziell: prismatische Formen, grüne Rampen, gezirkelt parallele Höhenschichten. Die Bewegungen der Böschungen, die feinen Modellierungen und die darauf abgestimmten Pflanzenmuster machen den Ring zu einem plastischen Raumereignis.

Eine formale Annäherung an den vorhandenen Berg wird nicht gesucht, er wird liebevoll umrahmt, herausgehoben in seiner ganz anderen Natur. Die schroffen Steinwände des früheren Steinbruches und die wilden Wäldchen an seinen Flanken werden inszeniert als Zeugnis einer anderen Zeit.

Auch die Inseln im Umfeld der Seen, die Themengärten, sind in ihren Umrissen und den immer gleichen geneigten Kanten erkennbar künstlich. Dennoch

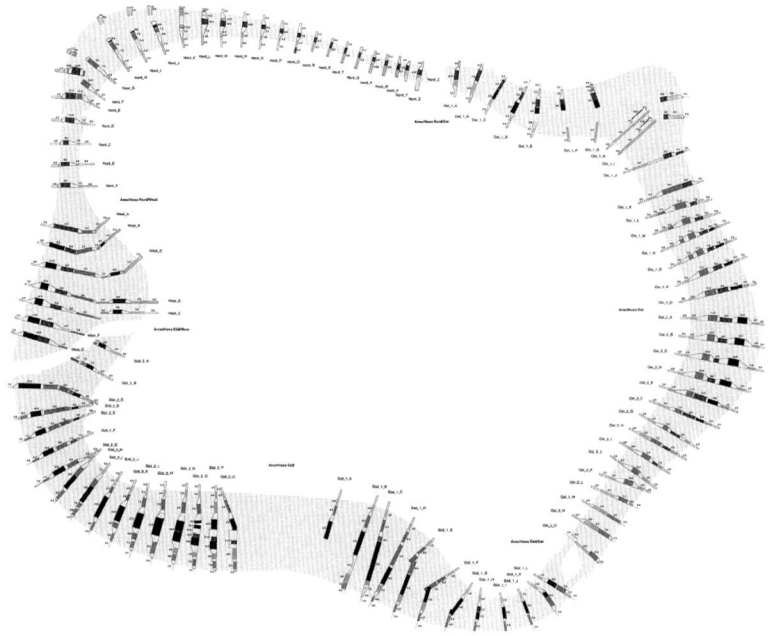

Schnitte durch den Ring, Übersichtsplan Section through the ring, ground plan

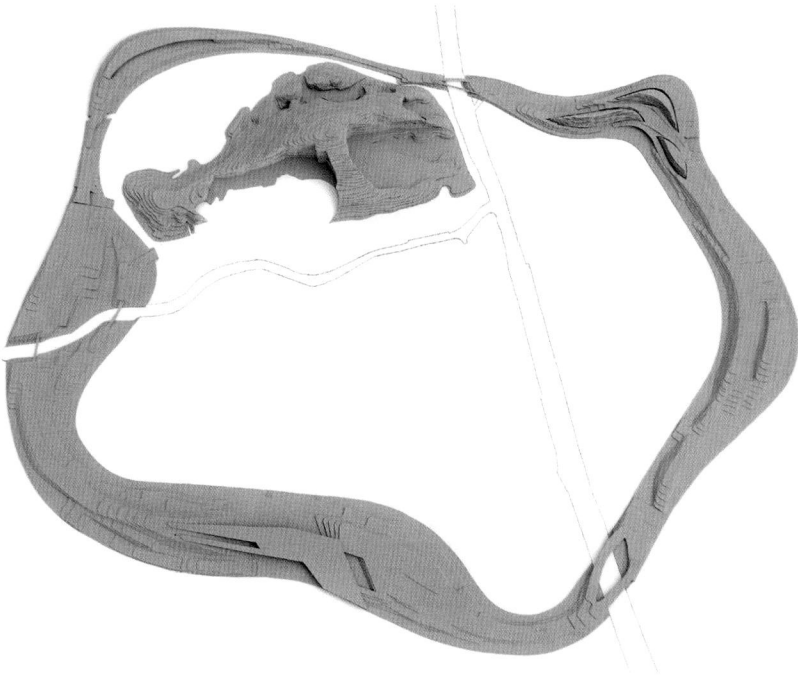

Schichtenmodell des Rings Topographical model of the ring

A formal attempt to converge with the existing hill has not been attempted, instead it is gently framed and its character, which is quite different, accentuated. The rough rock walls of the old stone quarry and the wild wooded areas on its flanks are presented as witnesses to another period of time.

Even the islands around the lake, the theme gardens, are noticeably artificial due to their shapes and similarly sloping edges. The landscape may actually have looked like this before the alluvial land silted up, an interplay of water, marshes and land with fast-changing boundaries and bizarre shapes.

Finding the new topography was an aesthetic and technical challenge. Grading a five-kilometre long embankment, which is up to 14 metres in height, is not something a landscape architect does on a daily basis. A variety of different methods was attempted, and immense plastiline models were built; the design was kneaded, in the truest sense of the word! The next step involved building a cardboard model, the accuracy of which was checked through the making of countless numbers of sections. In the end, the contour lines were developed in extremely detailed drawings. This great effort, this grappling with the form, was necessary in order to refine the design. It also convinced our Chinese partner, was respected, and strengthened everyone's resolve to continue with this experiment.

In Chenshan a large city does not yet exist. The visual focus is not yet defined, the aesthetic qualities are unknown. One thing is certain: the landscape that one now sees from the ring will disappear. For this reason an exclusion of the city was chosen instead of an attempt to merge with it, which is true to the archaic idea of the garden, the *hortus conclusus*. A view across the lake into the spaciousness of the garden of paradise is only possible upon crossing the embankment.

This is where a game of contradiction begins with the relationship between wilderness and garden. The artificiality of the scenery is accentuated by the grading of the site. The lines of the embankment, its slopes, and its interruptions are undoubtedly artificial: prismatic shapes, green ramps, and circling parallel layers of contours. The embankment's transformations, the subtle grading, and the coordinated patterns of plants all turn the ring into a sculptural experience.

A Waterscape The alluvial land of Chenshan was created by water. And water continues to characterize the image of this area. Heavy barges ride along navigable canals, and will ride through the garden in the future, and there are brooks, streams, and geometrically formed fish ponds. Water is noticeable everywhere, even underfoot; the entire landscape breathes humidity. And yet, the water causes no

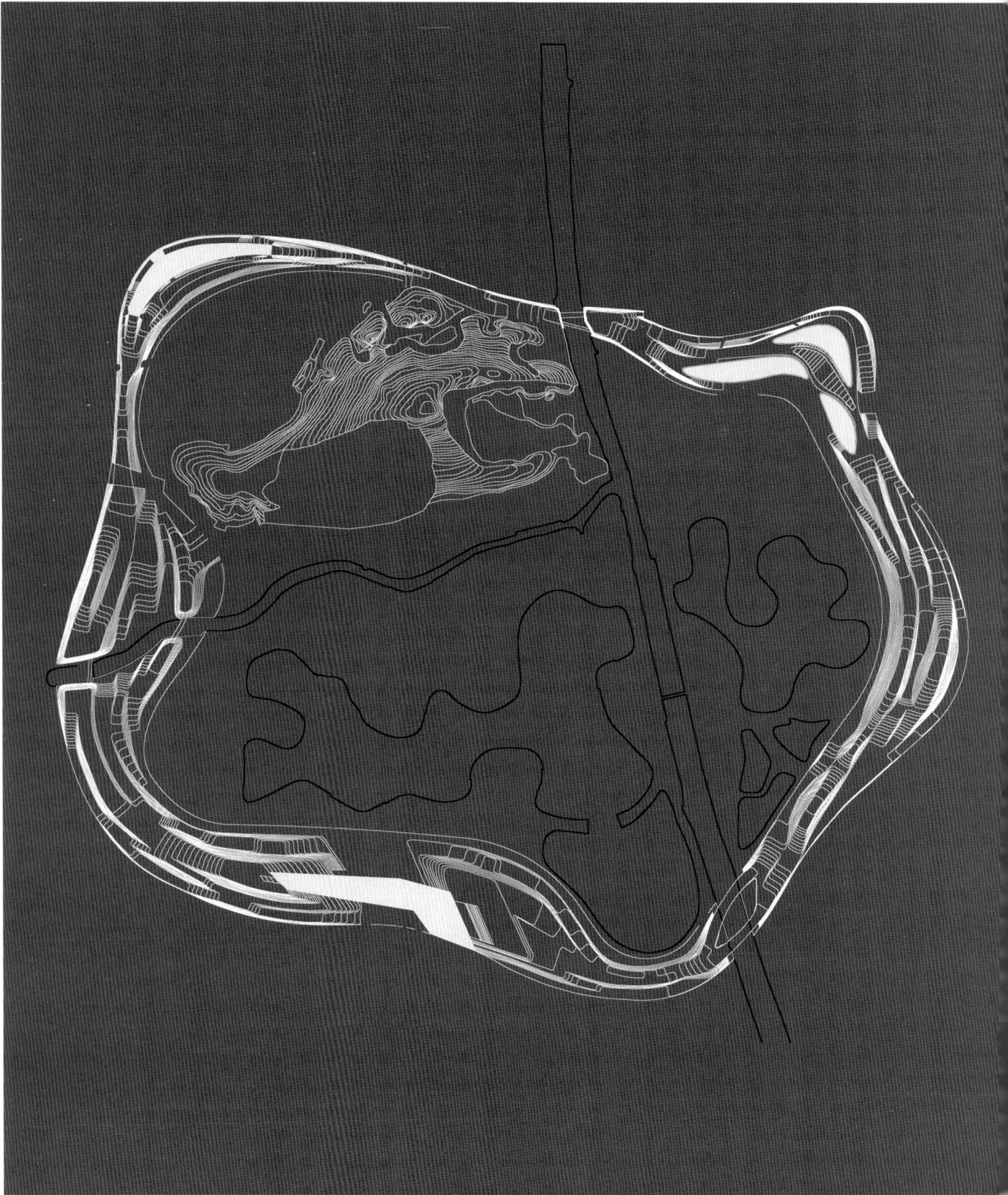

„Verbindung von Himmel und Erde mit einer
magischen Linie
Die Einheit von Himmel und Erde, von Natur
und Mensch, ist einer der zentralen Inhalte
der taoistischen Philosophie. Darum muss
dieses Thema auch in die Gestalt der Stadt
integriert werden. Aufmerksam gemacht auf
dieses Thema wird mit virtuellen, aber auch
im Raum lesbaren Linien, die von einem
„himmlischen" Ort, meistens einer Berg- oder
Hügelspitze, hin zur Stadt verlaufen."
aus: Carl Fingerhuth, Learning from China, Das Tao der
Stadt, Birkhäuser, Basel, 2004, S. 108
Alfred Schinz, The Magic Square, Cities in ancient
China, Edition Axel Menges, Stuttgart/London, S. 416

The connection of heaven and earth with a
magic line
"The unity of heaven and earth, of nature and
people, is one of the central ideas of Taoist
philosophy. For this reason this theme has
to be integrated into the design of the city.
Attention is drawn to this theme with a virtual
line, which is also visually distinguishable,
that runs from a 'heavenly' place, usually a
mountain or hilltop, down to the city."
from: Carl Fingerhuth, Learning from China, Das Tao
der Stadt, Birkhäuser, Basel, 2004, p. 108
Alfred Schinz, The Magic Square, Cities in ancient
China, Edition Axel Menges, Stuttgart/London, p. 416

mag die Landschaft, ehe das Schwemmland sich festigte, einmal ähnlich ausgesehen haben, ein Spiel von Wasser, Sumpf und Land mit fließenden Grenzen und bizarren Formen.

Die neue Topografie zu finden, war eine ästhetische und technische Herausforderung. Die Modellierung eines fünf Kilometer langen, bis zu 14 Meter hohen Reliefs gehört nicht zu den täglichen Aufgaben eines Landschaftsarchitekten. Verschiedenste Wege der Annäherung wurden erprobt, ungeheure Plastilinmodelle entstanden, der Entwurf wurde geknetet, im wahrsten Sinne des Wortes, in der nächsten Stufe in Karton geschnitten, durch zahllose Querschnitte überprüft, schließlich in minutiöser Darstellung in Höhenschichtlinien fixiert. Diese große Mühe, dieses Ringen um die Form war nötig, um den Entwurf zu präzisieren. Aber es hat wohl auch die chinesischen Partner überzeugt, hat Respekt gefunden und den Mut gestärkt, dieses Experiment zu wagen.

Eine Wasserlandschaft Das Schwemmland von Chenshan ist durch Wasser entstanden. Und Wasser bestimmt das Bild bis heute: auf schiffbaren Kanälen werden schwere Lastkähne auch künftig durch den Garten fahren, dazu Bäche und kleine Flussläufe, geometrisch angelegte Fischteiche. Das Wasser ist überall spürbar, auch unter den Füßen, die ganze Landschaft atmet Feuchtigkeit. Aber es ist Wasser, das keine Freude macht, sichtbar missbraucht, trübe und schmutzig, wie häufig in China.

Dagegen stand der Wunsch der chinesischen Partner, im Mittelpunkt des Gartens große Seen entstehen zu lassen, mit verschwiegenen Buchten, Inseln, Lotusblumen und Trauerweiden, die sich im Wind wiegen. Es ist der romantische Traum, ohne den ein gelungener chinesischer Garten ganz offensichtlich nicht zu denken ist. Aber auch uns schien es reizvoll,

dieses Bild in die Mitte der rationalen, eher strengen Form des Walles zu stellen.

Schwierig war es, die technische Umsetzung zu entwickeln. Die Seen und die kleinen sie speisenden Flüsschen sollten sauberer werden, klarer und frischer. Und dies sollte mit biotechnischen Mitteln erreicht werden. Zwar werden die Seen aus Grundwasser gespeist, aber auch dieses ist durch Eutrophierung belastet. Nun wird ein Teil des Shengjing-Flusses abgezweigt und in einer Schilfkläranlage in mehreren Becken gereinigt. Danach fließt das Wasser wieder in das ursprüngliche Bett des Flusses im Zentrum des Botanischen Gartens, vom Kanal und den umlaufenden Gewässern nun durch Sperren abgetrennt. Über mit Schilf bewachsene Gräben wird das Wasser verteilt und in den großen See geleitet, der mit seinen bewachsenen Uferzonen ebenfalls zur Wasserreinigung beiträgt. Von dort fließt das klare Wasser über eine Schwelle in den letzten See, den Wassergarten, wo die empfindlichsten Pflanzenschönheiten kultiviert werden. Schließlich wird das Wasser dem großen Kanal als kleine Auffrischung zugeleitet.

Es ist ein kompliziertes System, ein Spiel mit kleinsten Höhenunterschieden, aber es sollte gelingen. Unterstützung bei der Ausführungsplanung fanden wir bei einem chinesischen Hydrologen, der in Deutschland studiert hatte und mit dem Wassersystem vor Ort ebenso vertraut war wie mit der Technologie der Schilfkläranlagen. Er machte diesen neuen Wasserkreislauf des Botanischen Gartens zu seiner Sache.

Die Welt der Pflanzen Man hätte sich einfachere Standorte denken können für einen Botanischen Garten. Hohe Grundwasserstände, salzhaltige Böden und kräftige Winde machen die Wuchsbedingungen im Schwemmland von Chenshan für viele Pflanzen

sense of joy, is visibly abused, and is murky and polluted, which is common in China.

Contrary to this situation was the Chinese partner's wish for large lakes to be created in the centre of the park, with hidden coves, islands, lotus flowers, and weeping willows that would sway in the wind. This is the romantic dream without which a successful Chinese garden cannot quite obviously do without. But the idea of placing this image in the middle of the rational, somewhat austere form of the embankment appealed greatly to us as well.

The technical realisation of the concept proved to be difficult. The lakes and small streams that fed them were supposed to be cleaner, clearer and unsullied. And this was supposed to be achieved through the use of biotechnical measures. Although the lakes are groundwater-fed, they are still polluted as a result of eutrophication. Now a section of the Shengjiang River is to be diverted and purified in a series of pools belonging to a constructed wetland.

Thereafter the water is to flow back into the original riverbed and into the centre of the garden, which is separated from the canal and surrounding water bodies by a series of barriers. The water will flow into channels filled with reeds and then be directed into the large lake, which itself has a vegetated shoreline that further contributes to the purification process. From here, the clean water flows over a verge into the last lake, a water garden, where the most sensitive vegetative beauties are cultivated. After this the water is reintroduced into the canal as a small infusion of new life.

This is a complex system, which involves a series of minute changes in elevation, but it should prove successful.

In final planning, we were supported by a Chinese hydrologist who had studied in Germany and was as familiar with the aquatic system on site as we were with the technology of constructed wetlands. He took on the responsibility for the Botanic Garden's new water circulation system.

The World of Plants There are unquestionably places where it is easier to construct a botanic garden. A high groundwater table, saline soils, and strong winds make it difficult for many plants to thrive in the alluvial land in Chenshan. Protected areas for sensitive rare plants are now being created within the ring. These special gardens will be on islands raised above the moist subsoil, thus making them dryer, and allowing the plants to be presented in an incomparably sumptuous way. Trees and shrubs find a great variety of different growing conditions on the ring: narrow damp valleys, exposed hilltops, sunny slopes, and soggy hollows; a kaleidoscope of local conditions.

Offering both wilderness and garden, a promise that can be kept in this large park with its varied possibilities.

The Ring An essential decision on the way to reaching this goal was reached in our discussions with the botanic advisor Dr. Grau. Instead of a collection of trees from around the world, as originally planned, the goal of creating proper local growing conditions and sustainability became central to the design. The pathway around the ring, which is quite long, now runs past all of the earth's continents and especially through those types of forest and vegetation whose local conditions correspond to those found in the Shanghai region, i.e. a humid subtropical climate with little variation in temperature. This deliberate restriction is combined with the hope that, despite this difficult location, healthy plant communities can be

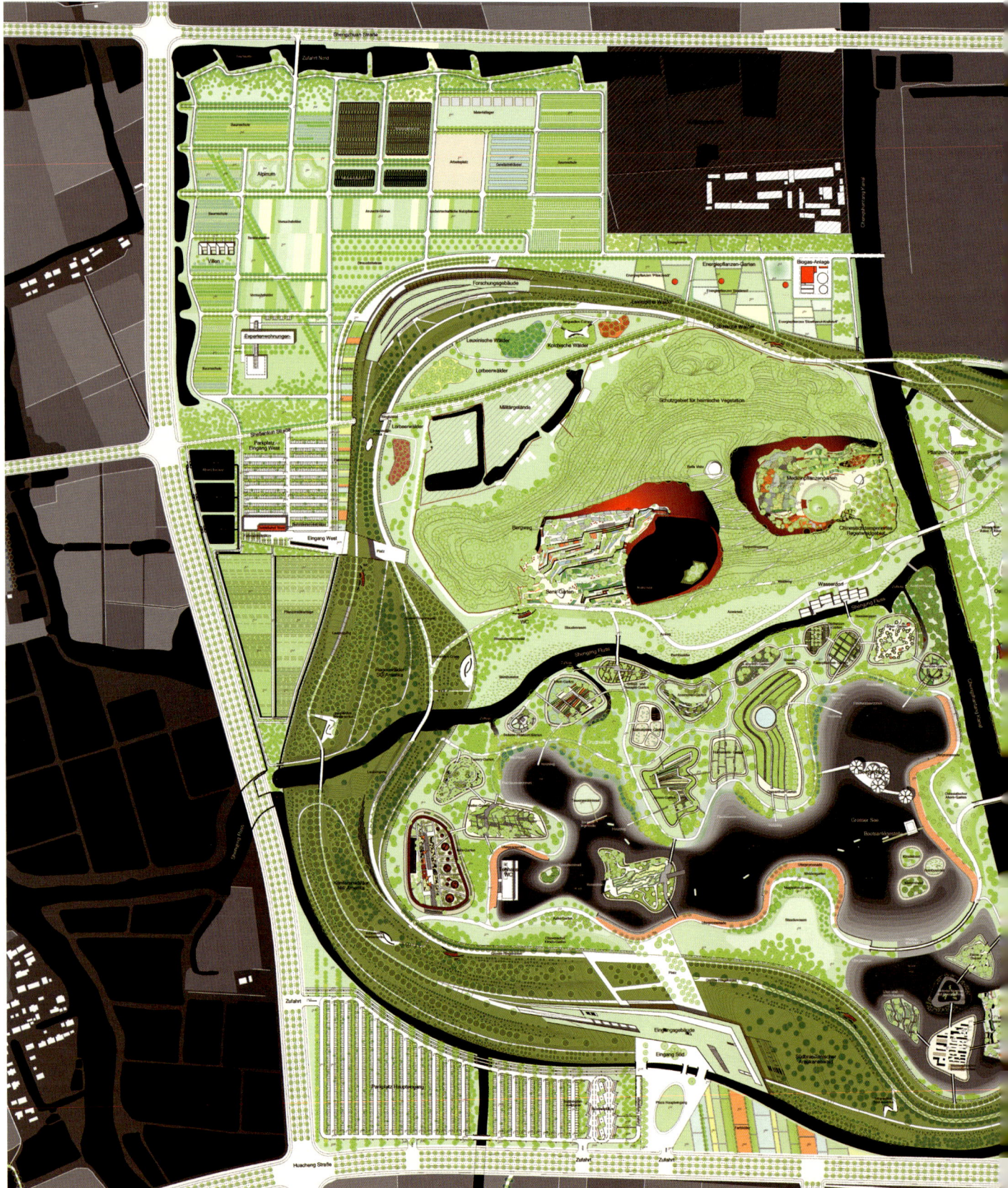

links: Masterplan Chenshan
Botanischer Garten Shanghai
left: Masterplan of the Chenshan
Botanic Garden, Shanghai
unten: Entwurfsstudie
below: Design study

Vegetationsgeografisches Konzept des Chenshan Botanischen Gartens Shanghai, mit Prof. Dr. Grau
Vegetative-geographic concept for the Chenshan Botanic Garden, Shanghai, with Dr. Grau

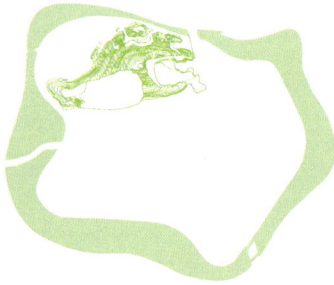

Präsentationsareal
Presentation Area

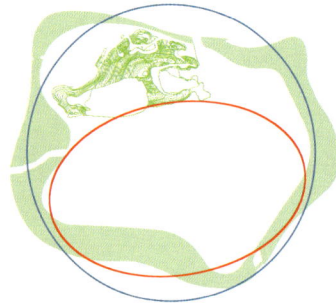

Zwei Routen
Two routes

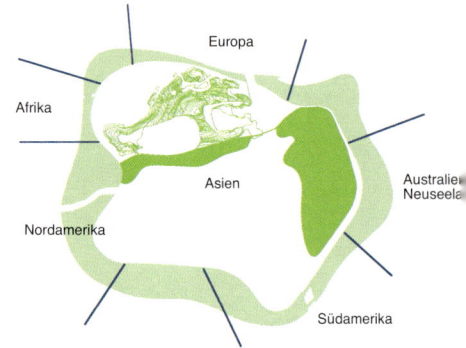

Europa
Afrika
Asien
Nordamerika
Australien Neuseela
Südamerika

Zonierung nach Kontinenten
Zoning according to continents

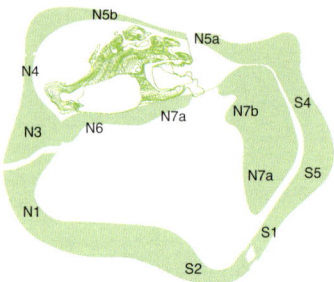

N5b N5a N4 N7b N3 N6 N7a S4 N1 N7a S5 S1 S2

Zonierung nach Vegetationsgebieten
Zoning according to vegetative zones

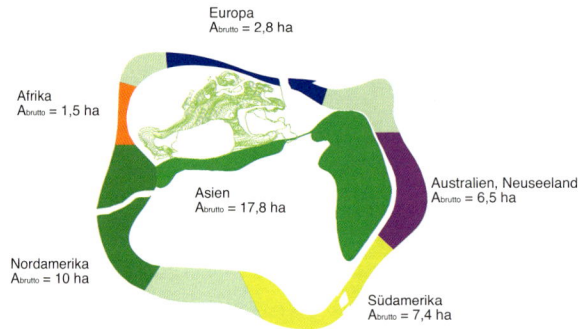

Europa A_{brutto} = 2,8 ha
Afrika A_{brutto} = 1,5 ha
Asien A_{brutto} = 17,8 ha
Australien, Neuseeland A_{brutto} = 6,5 ha
Nordamerika A_{brutto} = 10 ha
Südamerika A_{brutto} = 7,4 ha

Arealgrößen brutto
Area of site

Feinzonierung mit Leitbaumarten
Detailed zoning using trees

Zone N1 Koniferenwälder NW Nordamerika Coniferous forests of Northwest America
1 Nadelhölzer Conifers
Thuja plicata, Tsuga heterophylla, Picea sitchensis, Larix occidentalis, Abies grandis, Pinus monticola
2 Laubgehölze, Acer, Rhododendren Broad-leaved trees and shrubs, Maples, Rhododendron
3 Sequioa sempervirens (Redwood-Mammutbäume) auf tiefgründigen, sandigen Böden Redwood on deep and sandy soils

Zone N3 Regenwälder SO Amerika Rain forests of South-East America
1 Pinus, Quercus
sandiges Gelände Sandy soils
2 Taxodium (Sumpfzypressen Swamp Cypress) ins abfallende Wasser in riverbanks and wetlands
3 Magnolia, Liquidambar
nährstoffreiche Böden nutrient-rich soils

Zone N4 Makaronesischer Lorbeerwald Micronesian laurel forests
Lorbeerwälder der Kanaren und von Madeira Laurel forests on the Canary Islands and Madeira
1 Pinus canariensis
2 Laurus canariensis, Apollonias canariensis, Persea indica, Ocotea foetens, Ocotea bullata

Zone N5 Kolchische Wälder Cappadocian forests

Zone N5a Hyrkanische Wälder Hyrcanian forests
1 Zelkova caucasica, Diospyros lotus, Staphylaea colchica

Zone N5b Euxinische Wälder Euxinic forests
1 Tilia tomentosa, Arbutus andrachne, Laurus nobilis, Fagus orientalis, Rhododendron ponticum
2 Quercus iberica, Quercus frainetto

Zone N6 Südabfall des Himalaya Southern slopes of the Himalayas
1 Rhododendron arboreum, Rhododendron falconeri, Rhododendron grande, Rhododendron lindleyi, Cinnamomum camphora, Ilex-Arten, Ilex species, Camellia-Arten Camellia species
2 Castanopsis indica, Schima wallichii, Ilex-Arten, Camellia-Arten Ilex species, Camellia species

Zone N7a Chinesisches Temperiertes Regenwaldgebiet Chinese temperate rainforest area
1 Bestandsbäume Existing trees
2 Quercus acutissima, Eucommia ulmoides, Cunninghamia lanceolata, mehrere Magnolien-Arten several Magnolia species, Liquidambar formosana, Pinus armandii. Glyptostrobus lineatus

Zone N7b Regenwälder Japans Rainforests of
Japan
1 Castanopsis cuspidata, Machilus thunbergii,
Neolitsea aciculata, Lindera erythrocarpa,
Distylium racemosum, Cinnamomum japonicum,
Torreya nucifera, Chamaecyparis obtusa,
Cryptomeria Japonica, Sciadopytis verticillata

Zone S1 Valdivianischer Regenwald Valdivian
temperate rain forests
1 Fitzroya-Wälder Fitzroya forests
Fitzroya cupressoides
2 Aextoxicum-Wälder Aextoxicum forests
Aextoxicum punctatum, Eucryphia cordifolia,
Drimys winteri, Gevuina avellana, Lapageria rosea,
Nothofagus obliqua, Nothofagus procera
3 Nothofagus-Arten Nothofagus species
Nothofagus antarctica, Nothofagus pumilio,
Luma apiculata, Fuchsia magellanica, Gunnera
chilensis
(Humusreiche Böden, meist auf saurer Granite-
Basis, z.T. auch vulkanisch beeinflusste Böden)
(Humus-rich soils, usually on acidic granite
basis, also some volcanically influenced soils)

Zone S2 Südbrasilianischer Araukarienwald
Southern Brazilian Araucaria forests
1 Araucaria angustifolia, Cederla fissilis
Ilex paraguariensis, Phoebe porosa,
Berberis laurina, Dicksonia sellowiana (Baumfarn
tree ferns)
Relief relativ flach, z.T. auf Hochplateaus relatively
flat areas, partially on high plateaus
2 Podocarpus-lamberti-Wälder (in Tälern und
Senken) Podocarpus lamberti forests (in valleys
and depressions)

Zone S5 Neuseeländischer Regenwald New Zea-
land rainforests
1 Agathis-Wälder Agathis forests
Agathis australis, Beilschmiedia tawa,
Beilschmiedia taraira, Dacrydium cupressinum
2 Lorbeer-Koniferenwald Laurel coniferous forests
Rhopalostylis sapida, Metrosideros umbellata
3 Nothofagus-Wälder Nothofagus forests
Nothofagus menziesii, Nothofagus solandri,
Nothofagus truncata

Zone S4 Australischer Regenwald Australian
rainforests
1 Eucalyptus-Wald Eucalyptus forest
Eucalyptus calophylla, Eucalyptus delgatensis,
Eucalyptus megacarpa, Eucalyptus wandoo,
Eucalyptus diversicolor
(Humusreiche Böden auf Quarziten Humus-rich
soils on quartzites)

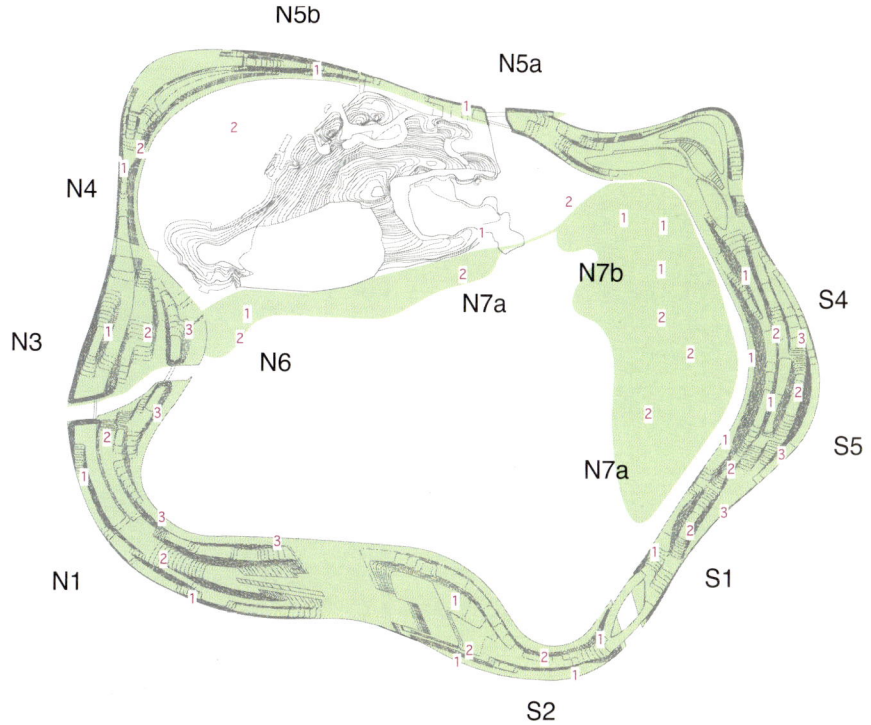

Lorbeerwälder – Zwei Routen durch die temperierten und subtropischen Wälder der Erde
Laurel forests – Two routes through the temperate and subtropical forests of the world

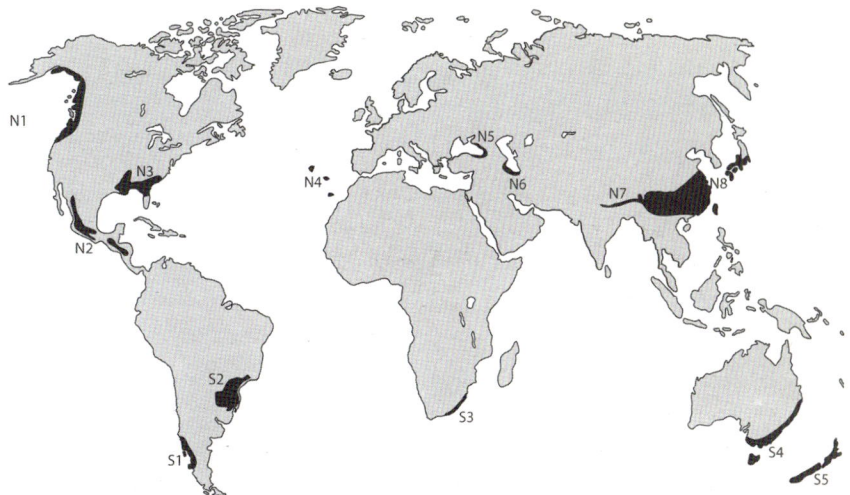

Weltweite Verteilung der temperierten und subtropischen Wälder der Erde
Distribution of the temperate and subtropical forests of the world

schwierig. Nun entstehen innerhalb des Ringes geschützte Standorte für die empfindlichen Raritäten. Die besonderen Gärten werden zudem Inseln gleichend aus dem feuchten Untergrund herausgehoben; sie werden dadurch trockener, die Pflanzen aber auch ungleich kostbarer präsentiert. Bäume und Sträucher finden auf der bewegten Landschaft des Ringes vielfältige Wuchsbedingungen: enge, feuchte Täler, windexponierte Kuppen, sonnige Hänge und feuchte Mulden, ein Kaleidoskop der Standortbedingungen.

Wildnis und Garten, ein Versprechen, das in diesem großen Park mit seinen vielfältigen Möglichkeiten eingelöst werden kann.

Der Ring Eine wesentliche Setzung auf dem Weg zu diesem Ziel wurde durch die Diskussionen mit dem beratenden Botaniker Prof. Dr. Grau gefunden. Statt einer Sammlung der Bäume der Welt, wie ursprünglich geplant, wurde das Ziel der Standortgerechtigkeit und der Nachhaltigkeit für den Entwurf bestimmend. Der Spaziergang um den Ring, ein langer Weg, führt nun durch alle Kontinente der Erde und dabei speziell durch diejenigen Wald- und Vegetationstypen, deren Standortansprüche denen der Region Shanghai entsprechen, also feuchtwarmen, subtropischen Klimaten mit wenig schwankenden Temperaturen. Mit dieser bewussten Beschränkung verbindet sich auch die Hoffnung, trotz des schwierigen Pflanzenstandortes vitale Pflanzengemeinschaften etablieren zu können und den Pflegeaufwand in Grenzen zu halten.

Der Weg über den Ring wird zu einer Wanderung durch die Lorbeerwälder der verschiedenen Kontinente. Lorbeerwälder gehören zu den ältesten Waldtypen unserer Erde. Sie sind sehr artenreich und zeichnen sich durch eine Vielzahl meist immergrüner Holzgewächse aus. Die gigantischen dunklen Redwood-Regenwälder an der Westküste von Amerika gehören ebenso dazu wie die immergrünen Laubwälder Chinas oder die Araukarienwälder Brasiliens. Häufig wird das Bild geprägt durch Lianen und Epiphyten an den Bäumen und durch eine Krautschicht aus Farnen, großblättrigen Kräutern, Bambusarten und Gräsern.

Das Klima von Shanghai erlaubt es, die wichtigsten Vertreter dieser außergewöhnlichen Vegetationsform als aneinandergereihte Kette von Beispielen darzustellen. Erstmals können Lorbeerwälder so im Vergleich gezeigt werden. Da sie vom Menschen seit längerer Zeit intensiv landwirtschaftlich genutzt werden und mancherorts nur noch als Relikte vorhanden sind, kann gleichzeitig auch etwas zum Erhalt dieser Arten beigetragen werden.

An besonders hervorgehobenen Punkten auf dem Wall werden jeweils verschieden gestaltete Informationspavillons stehen. Filme und Ausstellungen zeigen dort den jeweiligen Kontinent, seine Landschaften, die Pflanzen und Tiere, die ökologische Situation.

Der Waldring wird wissenschaftlichem Anspruch genügen, zugleich aber auch ein großzügiges und nutzbares Parkgelände sein. Dem natürlichen Vorkommen nachempfunden, entstehen kleine Wälder, offene Haine oder auch Gruppen großer Solitäre in lebhaftem Wechsel. Relativiert wird der Anschein von Naturnähe durch mächtige Alleen und Reihen, die mit exquisiten Baumarten Wege und Hangkanten betonen. Es sind Abgesandte des Gartens und der nahen Megastadt.

Überhängender Felsen. Europäischer Stich nach chinesischer Vorlage Overhanging rocks. European engraving based on a Chinese original

established while keeping the maintenance budget to a minimum.

The pathway along the ring will be a walk through the laurel forests of different continents. Laurel forests are among the oldest types of forest on our planet. They are very rich in the number of species they contain, and distinguish themselves through a large variety of mostly woody evergreen plants. The gigantic dark redwood rainforests on the west coast of America belong to this type of forest as do the evergreen deciduous forests in China or the Araucaria forests of Brazil. The common image of these forests often has vines and epiphytes on the trees, an herbal layer of ferns, large-leaf herbaceous plants, bamboo, and grasses.

Shanghai's climate allows for the creation of a chain of examples of the most important representatives of this extraordinary type of vegetation. This is the first time laurel forests can be shown together in this manner. The fact that the land they grow on has often been used in intensive agriculture for a long time and that in many areas these forests exist only as relicts also means that this exhibition can contribute to the preservation of these plants.

Information pavilions of varying design will be constructed at locations of special interest along the em-

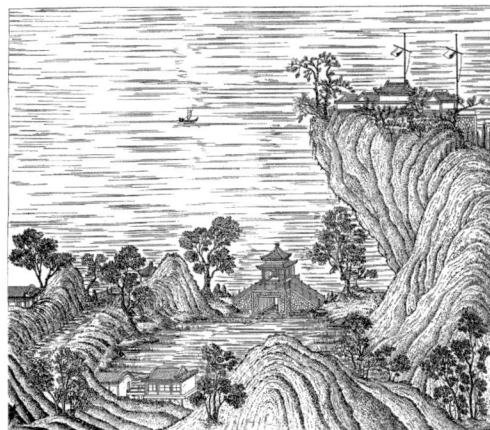

bankment. Films and exhibitions about a particular continent, a landscape, its plants and animals, and its ecological situation, will be shown at these points.

The forested ring will satisfy scientific requirements while at the same time remaining a generous and usable park. Modelled on a naturally occurring environment, small areas of woodland frequently alternate with open groves and groups of individual trees. This appearance of nature is relativized through the presence of impressive tree-line avenues and rows of trees, which emphasise paths and slope edges with exquisite types of trees. These represent emissaries of the garden and the nearby megacity.

N5 Kolchischer Wald

N7b Temperierte
Regenwälder Japans

N5b Hyrkanischer Wald

N5a Euxinischer Wald

4 Makaronesischer
orbeerwald

N7a Temperierte
Regenwälder Chinas

3 Immergrüne
Wälder des SO
Nordamerikas

N6 temperierte Regenwälder
des Himalaja

N7a Temperierte
Regenwälder Chinas

N1 Koniferenwald
des NW Nordamerikas

S2 Brasilianischer
Araukarienwald

S4 Australische
Regenwälder

S5 Neuseeländische
Regenwälder

Valdivianischer Regenwald

1	Trachycarpus fortunei	25	Zelkova caucasica
2	Platanus hispanica geschnitten		Diospyros lotus
3	Gingko biloba		Staphylea colchica
4	Prunus persica	26	Fagus orientalis
	Malus spectabilis	27	Quercus freinetto
	Malus hupehensis	28	Salix babylonica
	Malus x micromalus	29	Magnolia salicifolia
	Prunus serrulata	30	Chamaecyparis obtusa
5	Liriodendron tulipifera	31	Magnolia spec.
6	Acer macrophyllum	32	Cinnamomum japonicum
7	Acer macrophyllum	33	Liquidambar formosana
	Rhododendron californica	34	Punica granatum
	Pseudotsuga menziesii	35	Pinus parviflora
8	Thuja plicata		Pinus bangiana
	Tsuga heterophylla	36	Prunus persica
	Picea sitchensis		Malus spectabilis
	Larix occidentalis		Malus hupehensis
	Abies grandis		Malus x micromalus
9	Larix occidentalis		Prunus serrulata
10	Campsis grandiflora	37	Eucalyptus calophylla
	Lonicera japonica		Eucalyptus delgatensis
11	Sequoia sempervirens		Eucalyptus megacarpa
12	Quercus fellox		Eucalyptus wandoo
	Quercus virginiana		Eucalyptus diversicolor
	Magnolia grandiflora	38	Nothofagus pumilio
13	Liquidambar styraciflua	39	Nothofagus antarctica
14	Wisteria sinensis	40	Nothofagus obliqua
	Wisteria floribunda		Nothofagus procera
15	Quercus virginiana	41	Magnolia grandiflora
	Quercus fellox	42	Clematis montana
	Pinus taeda		Clematis florida
16	Taxodium distichum		Vitis vinifera
17	Rhododendren:	43	Acacia dealbata
	Rhododendron arboreum		Acacia decurrens
	Rhododendron falconeri	44	Eucalyptus
	Rhododendron grande		Eucalyptus
	Rhododendron lindleyi	45	Morus alba
18	Gingko biloba	46	Diospyros kaki
19	Magnolia dediolata	47	Firmiana simplex
20	Pinus canariensis	48	Podocarpus lambertii
21	Celtis africana		Phoebe porosa
	Persea indica	49	Ilex paraguariensis
22	Tilia tomentosa	50	Araucaria angustifolia
23	Laurus canariensis		Dicksonia lawsonii
	Laurus nobilis	51	Salix babylonica
24	Quercus iberica		
	Quercus freinetto		
	Tilia tomentosa		
	Laurus nobilis		
	Fagus orientalis		
	Rhododendron ponticum		

Anemonengarten Anemone Garden

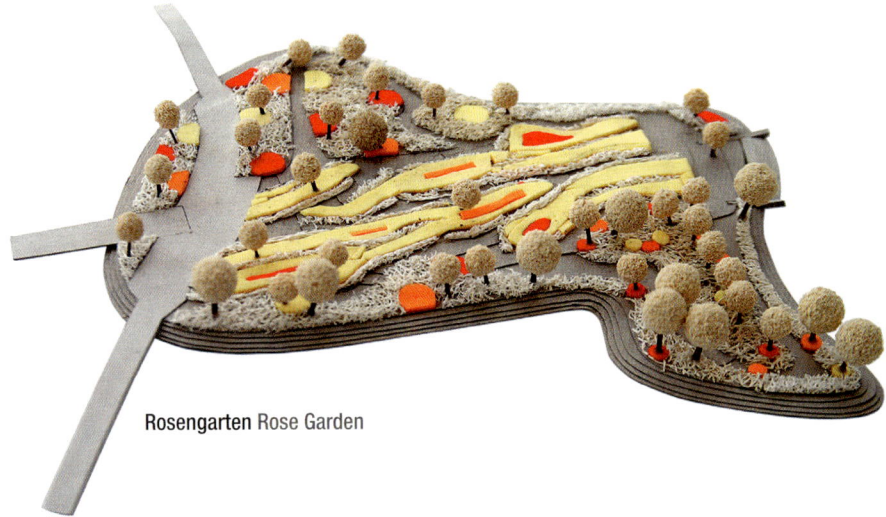

Rosengarten Rose Garden

Die Themengärten Die mehr als 35 Themengärten liegen wie Inseln in einer weit verzweigten Deltalandschaft. Sie folgen einer einheitlichen Grundstruktur. Etwa einen Meter aus dem feuchten Untergrund herausgehoben, gerahmt durch Böschungen mit einheitlichen Neigungen, durch Naturstein befestigt, bilden die Umrahmungen ein strenges und wiederkehrendes Muster in der weichen Umgebung. Individualität zeigt sich auf den Inseln. Dort wird der Garten gefeiert und gezeigt, welche gärtnerische Vielfalt und Pracht in den Jahrhunderten der Kultivierung von Pflanzen entstanden ist.

Jede der Inseln ist eine eigene Welt mit eigener Atmosphäre und individueller Gestaltung im Dienst der Pflanzenschönheit, die dort präsentiert wird: der Rosengarten, der Irisgarten, der Liliengarten, der Päoniengarten, der Topiarygarten und viele mehr.

Im Schönen wird Nützliches gezeigt. Ein Garten in Zuordnung zur Biogasanlage demonstriert die Bedeutung der Energiepflanzen, ein anderer zeigt Pflanzen, die zur Gewinnung von Fasern genutzt werden wie zum Beispiel Kokos, Hanf, Sisal oder Baumwolle. Eine Insel mit Olivenhain präsentiert die verschiedenen Ölpflanzen, eine andere Gemüsepflanzen in farbigen Mustern.

Eine besondere Rolle spielt, an einem ganz besonderen Standort, der große Medizinpflanzengarten. Im Mittelpunkt stehen die chinesischen Heilpflanzen; sie werden umrahmt von Heilpflanzengärten aus anderen Kontinenten wie dem Kräutergarten europäischer Klöster oder dem Waldheilgarten aus Sri Lanka.

Einige wenige Gärten sollen, *pars pro toto*, im Folgenden näher betrachtet werden:

Besondere Blüten Special flowers

The Theme Gardens More than 35 theme gardens lie like islands in a widely branching delta landscape. They adhere to a consistent basic structure. They rise approximately one metre above the damp subsoil, are surrounded by uniformly sloped banks, have natural stone paving, and form a framework around a rigid and repetitive pattern in the soft soil around them. Individuality is readily apparent on these islands. Here the garden is celebrated, and the variety and splendour resulting from the cultivation of plants over hundreds of years is displayed.

Each of the islands is a distinct world, and has its own atmosphere and individual design. Their function is to promote the vegetative beauty presented here: the rose garden, the iris garden, the topiary garden, and many more.

In beauty there is also utility. A garden allocated to the biogas plant demonstrates the importance of energy crops, another displays plants useful for the production of fibres, such as coconut palm, hemp, sisal, and cotton. An island with an olive grove presents different oil-producing plants, another shows vegetables in colourful patterns.

A large garden for medicinal plants, which has a special location, plays a unique role. Chinese medicinal plants are situated in the centre, surrounded by plants from other continents, such as herb gardens in European cloisters and medicinal forest gardens in Sri Lanka.

A few of the gardens should be looked at in more detail, *pars pro toto*:

60

Der Osmanthusgarten Dieser Garten ist einer in Europa wenig bekannten Pflanze gewidmet, dem Osmanthus, der Duftblüte. Sie wird hier in allen Farben und Variationen als Baum und Strauch gezeigt. Es sind immergrüne Gehölze mit betörendem Duft, deren Blüten vielfältig als Gewürz genutzt und auch dem grünen Tee als Duftmischung beigegeben werden.

Der Osmanthusbaum ist in der Legende mit dem chinesischen Mondfest verbunden. Die Nichte des Himmelsgottes wurde zur Strafe in einen Palast im Mond verbannt, begleitet nur von einem weißen Hasen, einer Kröte und dem duftenden Osmanthusbaum. In jedem Jahr wird zur Zeit der Osmanthus-Herbstblüte zur Erinnerung an diese Legende ein Fest gefeiert – nun auf dem kreisrunden Platz des Mondes, inmitten von Duftwolken im größten Osmanthusgarten Chinas.

oben: Osmanthusgarten, Grundriss
above: Osmanthus Garden, ground plan

The Osmanthus Garden This garden is dedicated to a little-known plant in Europe, the flowering Osmanthus. The plant is shown in all its variations, from tree to shrub. Osmanthus is an evergreen plant with a beguiling fragrance and flowers which are used in diverse spices and added to green tea as a scent.

The Osmanthus tree is associated with the legend of the Chinese Moon Festival. The Emperor of the Heaven's niece was banned to a palace in the moon as a punishment, and was accompanied by a white hare, a toad, and a fragrant Osmanthus tree. Every autumn, when the Osmanthus flowers, a festival is celebrated in remembrance of this legend – and now on the round Moon Plaza, surrounded by a heady scent in China's largest Osmanthus garden.

oben: Osmanthusgarten, Schnitte und Details above: Osmanthus Garden, sections and details

Der Labyrinthgarten Labyrinthe sind Archetypen, in aller Welt seit prähistorischer Zeit bekannt und in verschiedensten Formen gestaltet: in den Boden geritzt oder mit Steinen gelegt für kultische Tänze, als Grundriss ägyptischer Königsgräber und minoischer Paläste. Seit dem 17. Jahrhundert sind Labyrinthe ein in Europa sehr beliebtes Gartenmotiv, das der Klarheit und Ruhe des barocken Gartens ein sorgfältig geplantes Verwirrspiel gegenüberstellt. Meist ist der Grundriss mit immergrünen Gehölzen geometrisch bepflanzt, Hecken oder hohe Bosketts formen Mauern, manchmal bilden auch berankte Spaliere die Wände des Labyrinths. Wege führen den Besucher kunstvoll hindurch, sie sind das Sinnbild menschlicher Irrwege mit tröstlichem Ausgang, der Erlösung.

Für Chenshan wurde nicht die geometrische Form des Irrgartens gewählt. Vielmehr gleicht der Grundriss den Bildern, die durch kunstvolles Rechen des Kiesbodens in den japanischen Gärten entstehen. Die Wände bestehen aus unregelmäßig breiten Hecken des Wohlriechenden Schneeballs, der Boden ist richtungslos mit Wildpflaster belegt. Drei besondere Punkte erreicht man auf den verschlungenen Wegen: aus dem gleichmäßig hohen Labyrinth ragt eine hohe schlanke Halle, aus Bambusstäben gebaut und mit Bougainvillea berankt, an anderer Stelle ein Aussichtsturm, auch er ist aus Bambus gebaut und berankt mit blauen Glyzinien, und schließlich eine dunkle, feuchte Halle aus lebendem Bambus, in deren Mitte ein Nebelbrunnen eine geheimnisvolle Stimmung schafft. Den Rahmen um das Labyrinth bildet ein lichter Saum chinesischer Schirmbäume und Japanischer Zeder.

Labyrinthgarten, Grundriss
Labyrinth Garden, ground plan

The Labyrinth Garden Labyrinths are archetypes, known around the world since prehistoric times and found in a variety of styles: etched into the ground or formed with stones for Celtic dances, and as floor plans for the tombs of Egyptian kings and Minoan palaces. Labyrinths have been a favourite garden motif in Europe since the 17th century, confronting the clarity and peacefulness of baroque gardens with carefully planned confusion. Usually a ground plan is geometrically planted with evergreen shrubs. Hedges and high bosquets form walls and sometimes the walls of a labyrinth are formed by trellises and climbing plants. Paths skilfully lead visitors through this type of garden, which serves as a symbol of human aberration with a comforting way out, a form of deliverance.

The geometric form of labyrinth was not chosen for Chenshan. The ground plan is much more similar to the images created in a Japanese garden through the artistic raking of gravel. The walls consist of hedges of sweet-smelling snowball viburnams of irregular width, and the ground is covered with pavers installed in a directionless fashion. Three special points can be reached along the winding paths: rising above the labyrinth is a tall slender tower of bamboo stalks covered with bougainvillea, an observation tower also built of bamboo and covered with blue wisteria, and finally a dark, damp hall made of living bamboo, which has a misting fountain at its centre, creating a mysterious atmosphere. A light screen of Chinese parasol trees and Japanese cedars surrounds the labyrinth.

Der Wassergarten Wasser ist die Seele jedes chinesischen Gartens. Die großen Wasserspiegel geben Ruhe und Bewegung, Weite und Zeitlosigkeit und reflektieren den Himmel.

Die große, in den Ring gebettete Halle des Haupteinganges öffnet sich auf eine besonnte Promenade am Südrand der Seen. Mit Treppen, Stufen und Mauern wird der Blick inszeniert auf die vielfältig gegliederte Wasserlandschaft und die dahinter liegenden Zeugenberge.

Der architektonisch anmutende Wassergarten zeigt die wissenschaftliche Präsentation der Wasserpflanzen. In Bassins werden Wassertemperatur, Wassertiefe, Salzgehalt und Durchlaufgeschwindigkeit differenziert gesteuert, sodass auch seltene und empfindliche Arten gezeigt und unterschiedlichste Lebensgemeinschaften den Besuchern nahegebracht werden können.

Gegenüber – und im Kontrast dazu – liegt eine weite, naturnahe Wasserlandschaft mit flachen und steileren Ufern und allen Gradienten von Unterwasservegetation über Sumpf und Verlandungsbereiche bis hin zu den höher gelegenen trockeneren Inseln. Dort wandert man durch wilde, naturnahe Sumpf- und Uferpflanzen, unter Sumpfzypressen und Trauerweiden und entlang blühender Hochstaudenfluren.

Der Irisgarten senkt sich von der erhöhten Uferpromenade bis hin zum Wasser ab, ein System von Terrassen, Mauern und verborgenen Sitzplätzen. Die große Vielfalt der Irisarten und der Hemerocallis kann hier, vom trockenen bis zum sumpfigen Standort, ze-

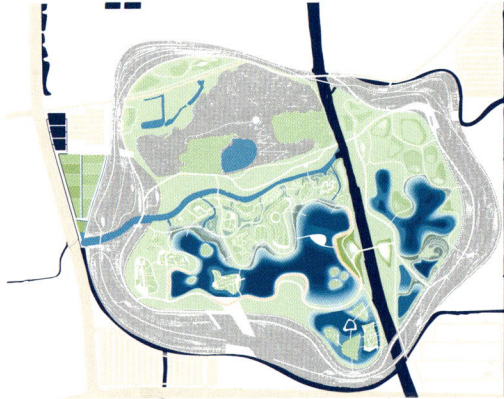

links: Wasserkonzept left: Water concept rechts: Übersichtsplan Wassergärten right: General plan of Water Gardens

lebriert werden. Es entfaltet sich vom zeitigen Frühjahr bis in den späten Herbst hinein ein wunderbares Farbenspektrum.

Schließlich durchquert man die Insel der Farne, eine eigene Welt. Auf fein modellierter Topografie steht ein dichtes Dach verschiedenartiger Bäume. In deren Schatten und auf verschieden feuchten Substraten wird die Vielfalt der Farne und Gräser demonstriert; als Kontrast zur farbigen Welt der Irisinsel findet sich hier nun Grün in allen Abstufungen und Strukturen, eine geheimnisvolle feuchte Atmosphäre im Halbschatten der Blätter und der riesenhaften Baumfarne.

Auf schwankenden Brücken und Stegen geht man dann über das Wasser, erlebt es ganz nah und spaziert zwischen Victoriablättern zu schwimmenden Seerosen und zu den Lotusfeldern – *lien-hua*. Die chinesische Lotusblüte und der Lotussee sind Synonyme für das Innere eines chinesischen Gartens.

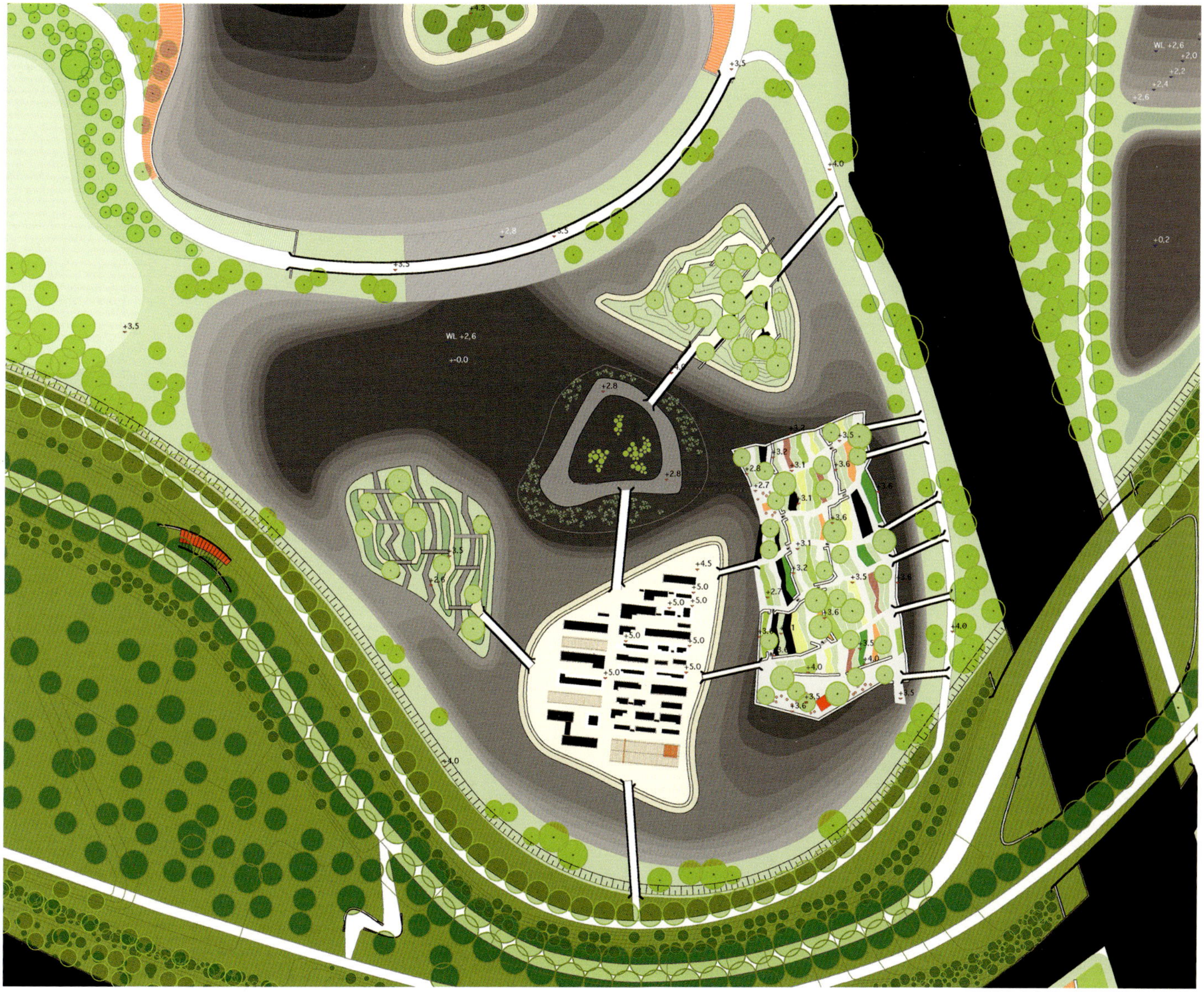
WL +2,6
+-0.0
+3.5
+2.8
+4.0
+4.5
+5.0
+0,2

The Water Garden Water is the soul of every Chinese garden. The water's extensive surface is serene, allows for movement, a sense of vastness and timelessness, and reflects the sky.

The large building at the main entrance, which is integrated into the ring, opens up to a sunny promenade at the southern edge of the lakes. From here the view is directed over stairways, steps, and walls to the waterscape's multifaceted structure and to the hill rising on the far side.

The architecturally styled water garden offers a scientific presentation of aquatic plants. Basins with varying water temperatures, depths, salinities, and speeds of flow have been created so that rare and sensitive species can be shown and different communities explained in detail to visitors.

Opposite this – and in contrast to it – is a broad natural water landscape with both flat and steep banks and all types of underwater vegetation ranging from swamps and silted-up areas to more elevated dryer islands. Here visitors can wander through wild and untouched swamp areas and riverbank vegetation, under bald cypress trees and weeping willows, and past blossoming stands of perennials.

The iris garden slopes away from the elevated promenade to the water's edge and consists of a system of terraces, walls, and concealed seating areas. A great variety of irises and Hemerocallus from diverse environments can be enjoyed here. A wonderful spectrum of colours is presented in this garden from early spring to the late autumn.

At the end of the garden visitors walk through the island of ferns, which is a world unto itself. The thick crowns of various types of trees are suspended over a subtly graded landscape. In their shadows a great range of ferns and grasses grow on different moist soils. Green in all its many shades and structures serves as a contrast to the colourful world of the iris island, creating a mysterious and sultry atmosphere in the half-shade of the leaves and the giant tree ferns.

Visitors cross the water on a variety of swinging bridges, remaining close to the water and walking between giant Victoria amazonica leaves and water lilies and then to the floating fields of lotus flowers, the *lien-hua*. Chinese lotus flowers and lotus-filled lakes are synonyms for the interior of a Chinese garden.

68

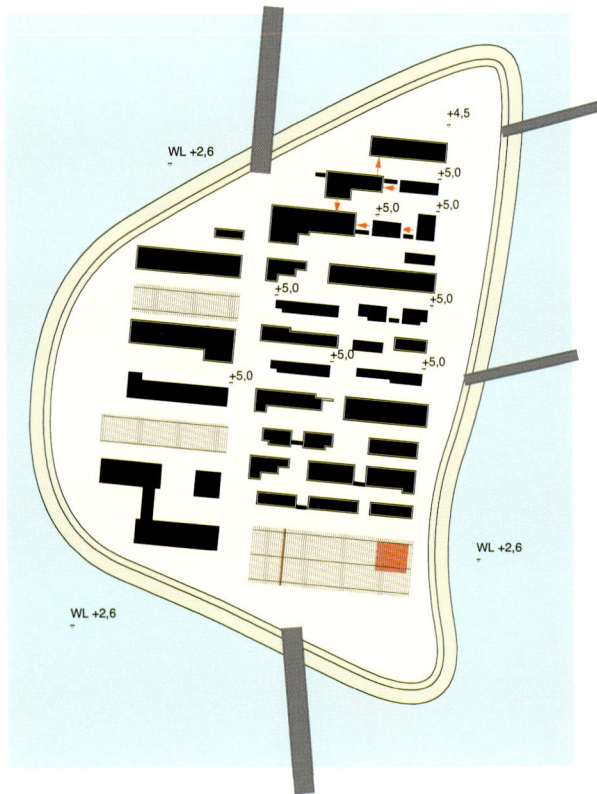

Pflanzengruppierungen mit schematischen
Mindestflächen Groups of plants with
smallest schematic areas

Feuchtflächen
Wetland areas

Sauergräser
Sedges

Besonders hervorzuhebende
Pflanzen
Plants to be accentuated

Submerse Pflanzen
Aquatic plants

Farne, Schachtelhalme, Algen
Ferns, rushes, algae

Seichte Becken
Shallow basins

Schlammkriecher
Bristleworms

Brackwasserpflanzen
Brackish water plants

Fleischfressende Pflanzen
Carniverous plants

Arundo donax
Arundo donax

Gunnera und Verwandte
Gunnera and related plants

Wassergarten Besonderheiten. Grundriss und Pflanzengruppierungen
Water Garden, special features. Ground plan and groups of plants

Seerosengarten, Grundriss
Water Lily Garden, ground plan

Farngarten, Grundriss
Fern Garden, ground plan

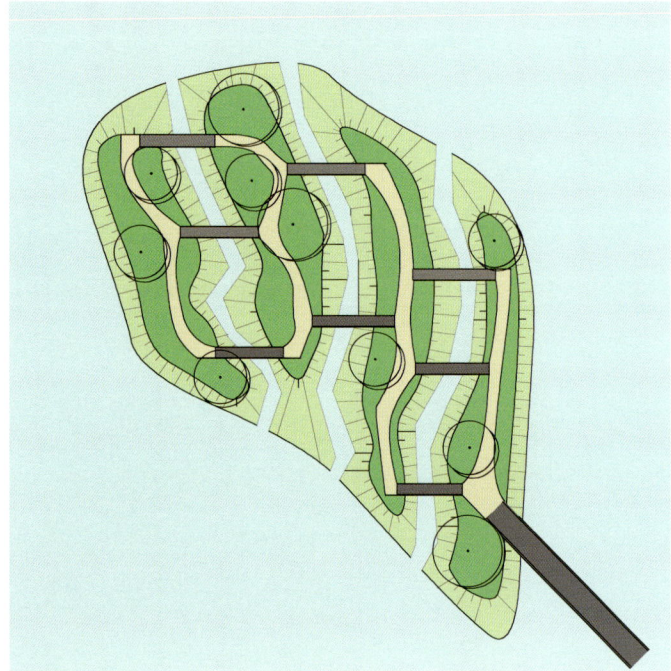

Ufer- und Hochstaudengarten,
Grundriss Riparian and Tall Herb
Garden, ground plan

Iris für trockene Standorte, wie z.B.
Irises for dry locations, e.g.,

- Iris aphylla
- Iris barbata
- Iris foetidissima
- Iris germanica
- Iris graminea
- Iris pallida
- Iris lusitanica
- Iris nigricans
- Iris pumila
- Iris spuria
- Iris tingitana
- Iris variegata

Iris für feuchte Standorte
Irises for moist locations

- Iris graminea
- Iris kaempferi
- Iris laevigata
- Iris pseudacorus
- Iris sibirica
- Iris versicolor

Zwiebel-Iris
Bulbous irises

- Iris anglica
- Iris bucharica
- Iris danfordiae
- Iris histroides
- Iris reticulata

Irisgarten, Grundriss
Iris Garden, ground plan

Zufahrt Nord

Militärgelände

Techhalle

Alpinum

Villen

Expertenwohnungen

Forschungsgebäude

Energiepflanzen-Garten

Biogas-Anlage

Leuxinische Wälder

Koreanische Wälder

Lorbeerwälder

Schutzgebiet für heimische Vegetation

Pflanzen-System

Chinesischer Regenwald

Bergweg

Shatunkun Straße

Eingang West

WC

Shenging Fluss

Regenwälder SO Amerika

Bambuswälder

Bruchwald

Grosser See

Bootsanlegestelle

Koniferenwald NW Amerika

Teehaus WC

Apfel-Garten

Chinesischer Kirsch-Garten

Magnolien-Garten

Staudensee

Eingangsgebäude

Südbrasilianischer Araukarienwald

Huacheng Straße

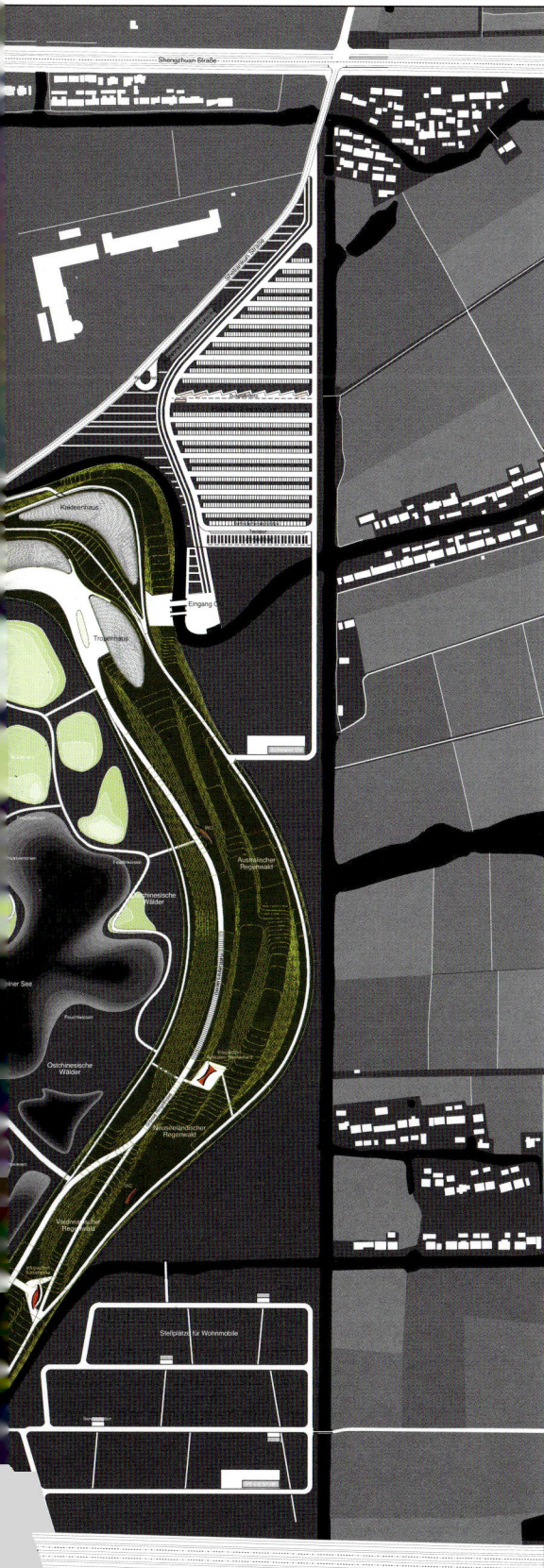

Infrastrukturplan
Infrastructure plan

FLÄCHENBILANZ INFRASTRUCTURE PLAN

Gesamtfläche	100%	ca. 200 ha
Grünflächen	62%	123 ha
Wasserflächen	17%	34 ha
Befestigte Flächen	18%	36 ha
Gebäudefläche	3%	5 ha

Bereiche	
Pflanzengeografie Ring	43 ha
37 Themengärten	20 ha
Forschungsgelände	30 ha

Gebäude	
Haupteingangsgebäude	20.700 m^2
Forschungsgebäude	17.900 m^2
Gewächshäuser	19.400 m^2

Parkplätze	
Hauptparkplatz:	1320 PKW
Parkplatz Ost:	990 PKW
Parkplatz West:	430 PKW

Total area	100%	ca. 200 ha
Vegetated areas	62%	123 ha
Water	17%	34 ha
Paved surfaces	18%	36 h
Buildings	3%	5 ha

Areas	
Plant-geographic section of Ring	43 ha
37 theme gardens	20 ha
Research area	30 ha

Buildings	
Main entrance building	20,700 m^2
Research building	17,900 m^2
Greenhouses	19,400 m^2

Parking areas	
Main parking area:	1320 cars
Parking area east:	990 cars
Parking area west:	430 cars

74

Die Architektur Die Hauptgebäude des architektonisch-baulichen Programms, das Eingangsgebäude, das Forschungszentrum und die Großgewächshäuser, sind in das Kontinuum des Gartenringes integriert.

Sie werden selbst Bestandteil des Ringes und damit der Landschaft. Sie überhöhen gleichsam die Idee des Gartens und bilden Orte der konzentrierten Erfahrung.

Mit ihren in Grund- wie Aufriss dynamischen Formen und der wechselnden Materialität zwischen Beton als Träger der Landschaftsebene und Glas als transparenter Füllung zwischen den Aufweitungen des Bandes, fügen sie sich wie selbstverständlich in die Vorgaben der Gartenarchitektur ein.

Auftakt des Botanischen Gartens ist das große, in den Wall geschobene Empfangsgebäude. Es ist Teil des Ringes, über seinem Dach schließt sich der Rundweg, und doch ist es markante Architektur. Breite Fensterbänder schneiden sich in die Flanken des Ringes und öffnen die Räume großzügig nach außen. Die Eingangshalle gleicht einer Schleuse zwischen der profanen Welt davor und dem inneren Garten. Großzügig öffnet sie sich nach Süden zur Plaza und nimmt den ankommenden Besucher auf. Hier findet er das Besucherzentrum mit verschiedenen Veranstaltungsräumen, Läden, Restaurants und wechselnden Ausstellungen. Ebenso großzügig weitet sich dann die Halle nach Norden und gibt den Blick frei auf die Seenlandschaft des Gartens.

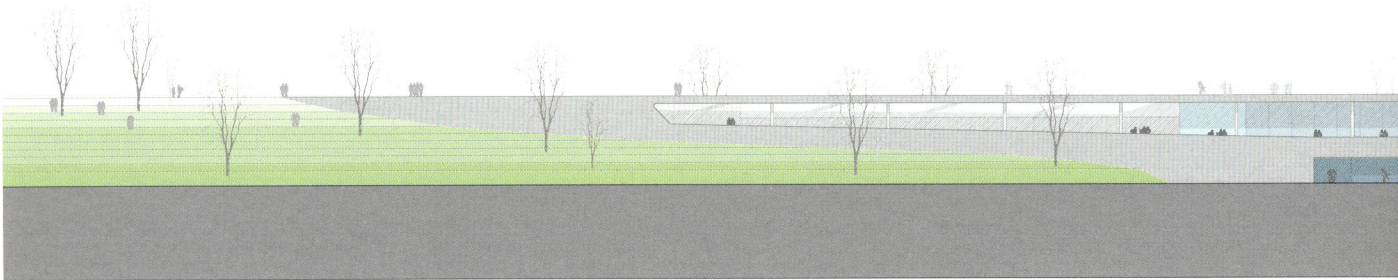

Eingangsgebäude, Ansicht Nord Entrance building, north elevation

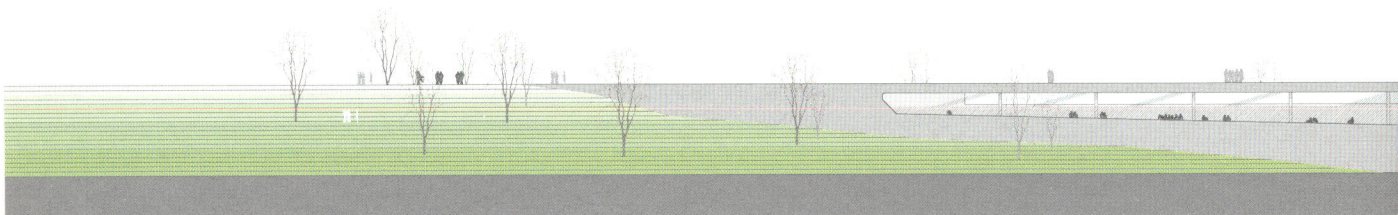

Eingangsgebäude, Ansicht Süd Entrance building, south elevation

The Architecture The architectural programme's main buildings, i.e. the entrance building, the research centre and the large green houses, are integrated into the continuum of the garden ring.

They become elements of the ring and thus of the landscape as well. In a way, they elevate the idea of the garden and create an area of concentrated experience.

With their dynamic ground plans and elevations and the varying materiality between concrete as a support at the level of the landscape and glass as a transparent filling where the ring broadens, they naturally merge with the prerequisites of the landscape architecture.

The Botanic Garden begins at the large entrance building, which is built into the embankment. It is an element in the ring, with the circular pathway running across its roof, and yet it remains a striking piece of architecture. Broad bands of windows cut into the flanks of the ring open the interior to the outside environment in a generous way. The entrance hall acts as a gateway between the mundane world on the outside and the interior garden. It broadly opens to the plaza on the south side and welcomes oncoming visitors. Here, the visitor centre offers a variety of rooms for events, stores, restaurants and temporary exhibitions. The hall also opens generously to the north, revealing a view of the garden's lakeland area.

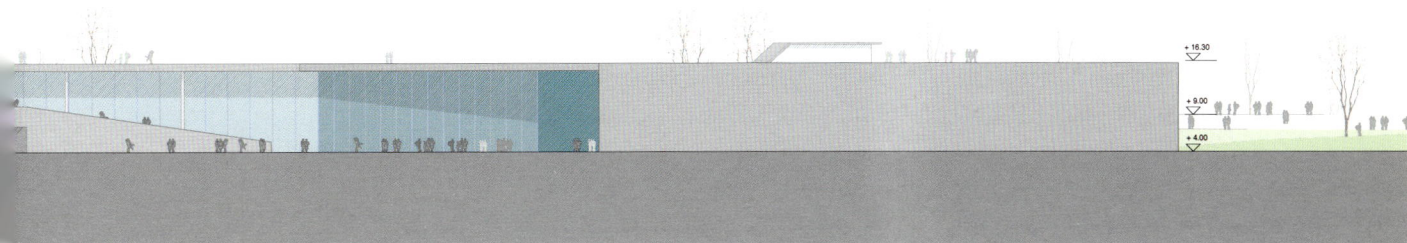

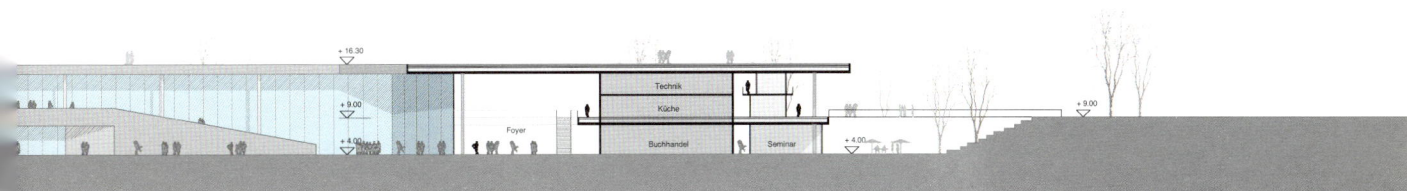

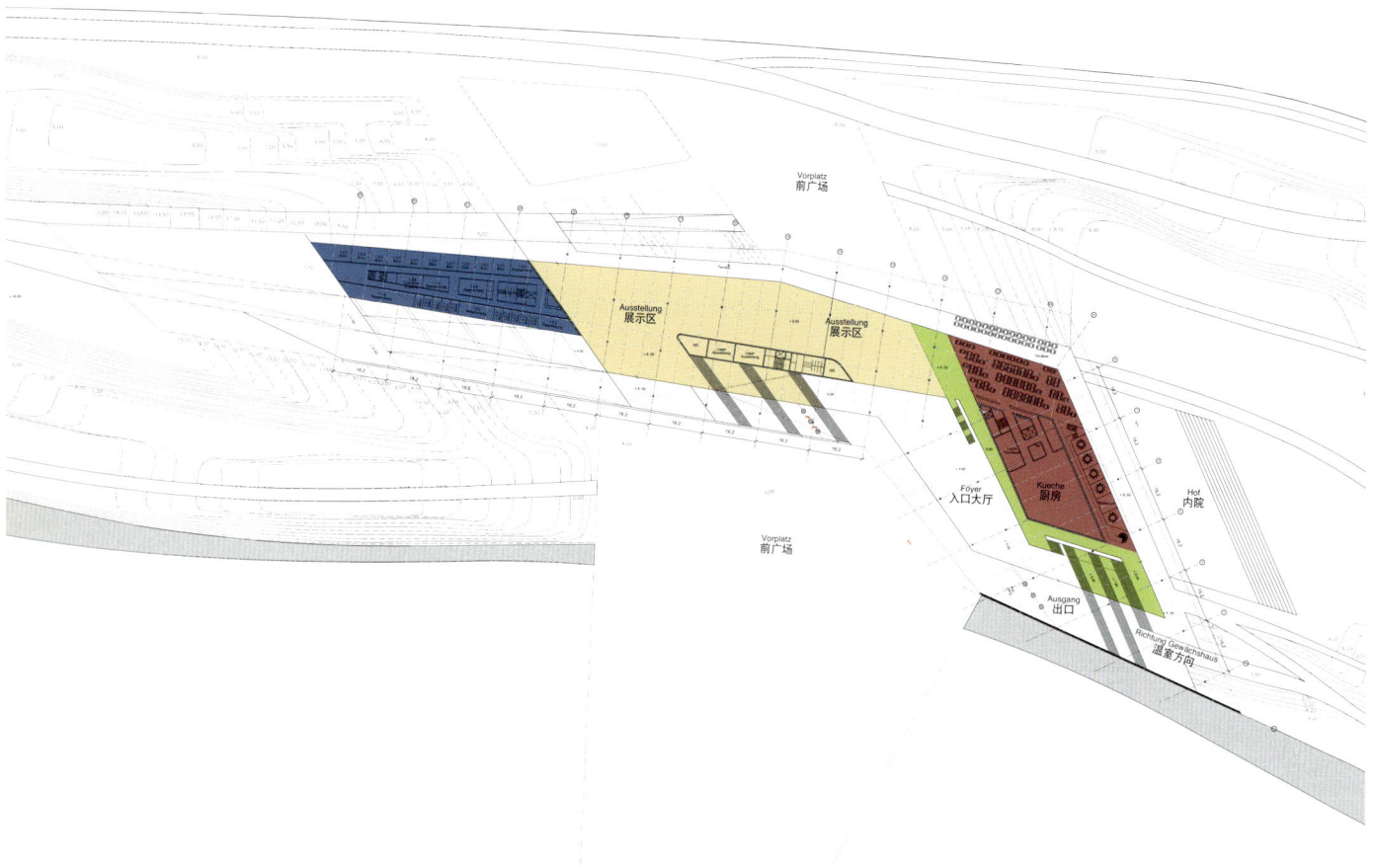

Eingangsgebäude, Erdgeschoss Entrance building, ground floor

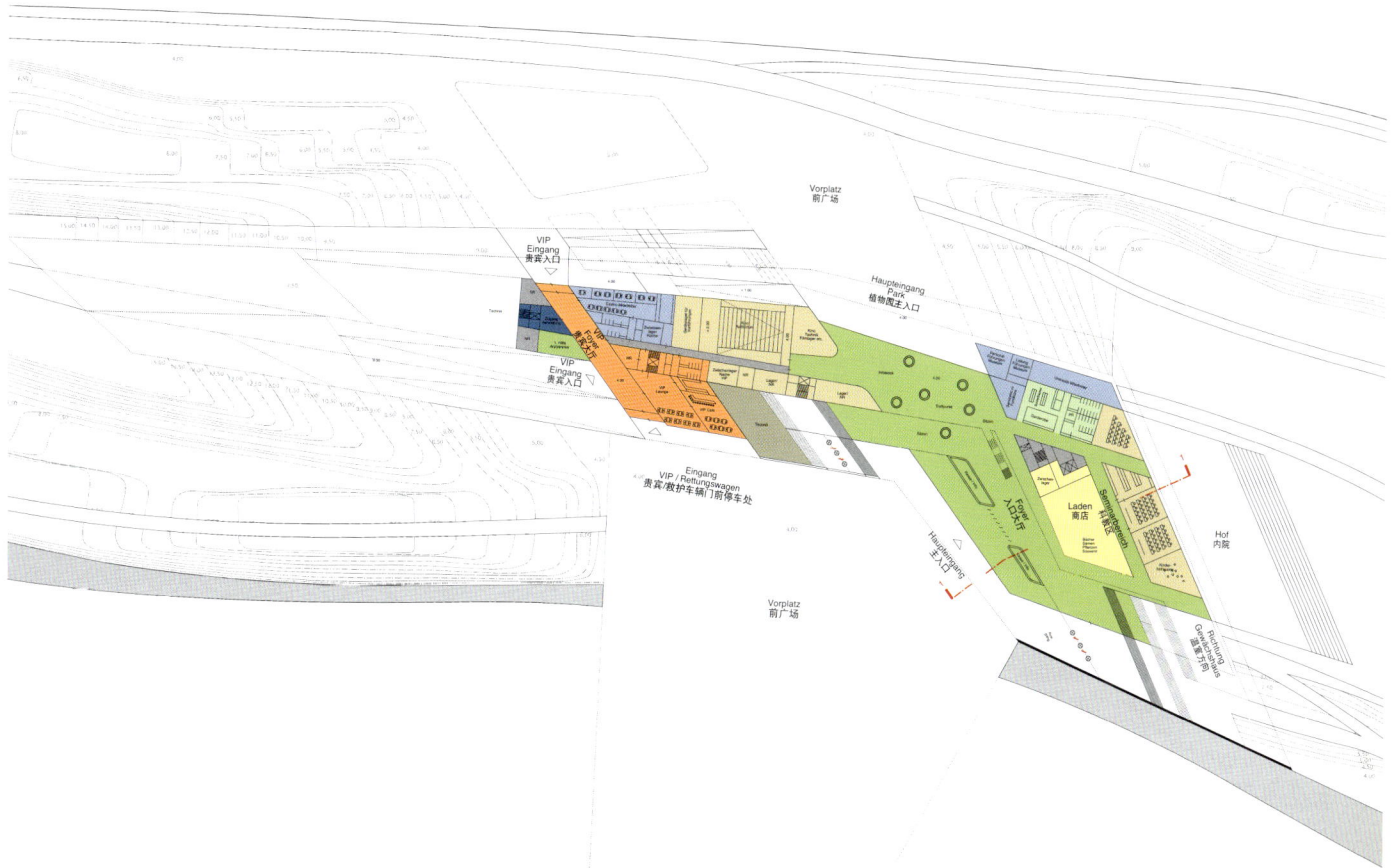

Eingangsgebäude, Obergeschoss Entrance building, upper floor

Nahe dem nordöstlichen Eingang und dem Shengjing-Fluss liegt ein zweiter baulicher Schwerpunkt: ein Ensemble dreier gläserner Volumina. Der Ring weitet sich hier zu einem Geflecht verschiedener Stränge, in die auf linsenförmigen Grundrissen die drei großen Gewächshäuser eingewoben wurden. Insektenaugen gleich schauen die länglich gebogenen Kuppeln aus der Erdmodellierung und übersetzen deren Schwere in gläsernen Raum.

Im Zentrum der drei Hallen, durch das auch der Hauptweg führt, liegen ein Hörsaal, eine Shoppingmall und ein luftiges großes Restaurant. Das daran anschließende große Palmenhaus kann als Festsaal genutzt und von Gästen gemietet werden. Die beiden anderen Gewächshäuser nehmen auf, was auch in Shanghai nur in kontrollierten Klimaten überleben kann: Das Tropenhaus zeigt die Vegetation der Urwälder und vor allem die Sammlung der Orchideen, das Kakteenhaus die Flora der südchinesischen Wüsten, aber auch anderer Trockenregionen der Erde.

An additional architectural focal point is found near the northeast entrance and the Shengjiang River: an ensemble of three glass volumes. At this point the ring broadens to a network of different strands which are woven into the lenticular ground plans of the three large greenhouses. Similar to an insect's eyes, the long curving domes peer out of the graded earth, transforming its solid weight into a crystalline space.

An auditorium, a shopping mall, and a large, airy restaurant are located in the centre of the three halls, through which the main pathway also leads. Connected to this is a large palm house, which can be used as a ballroom and rented by guests. The remaining two greenhouses contain those plants which require a controlled climate to survive in Shanghai: The tropical house exhibits jungle vegetation and, above all, a collection of orchids, and the cactus house contains flora from southern Chinese desert regions as well as vegetation from arid regions around the world.

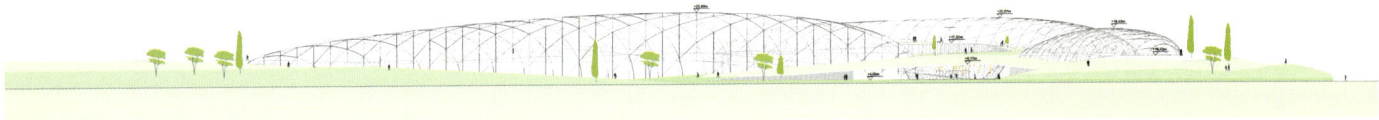

Gewächshäuser, Schnitte Greenhouses, sections

2.2.1.2.
Zufahrt
Geächshäuser
Anlieferung
运输入口

Kateen/Duenenpflanzen
仙人掌植物
1500 qm

Australien Pflanzen
澳洲植物
800 qm

4150qm

Afrika+Europa Pflanzen
非洲欧洲植物
750 qm

Hörsaal
报告厅

Gastronomie

Sukkulenten
多肉植物
1500 qm

Palmen Pflanzen
棕榈植物
1500 qm

Innengarten
室内花园
1500 qm

5400qm

Shop-Mall
商场

Foyer
共享空间

Haupteingang
主入口

Haupteingang
主入口

Orchideen
Ananaspflanzen
兰花凤梨植物
500 qm

Calathea Aroid
苏铁天南星植物
300 qm

2680qm

Schattenpflanzen
喜阴植物
(Baumfarne)
1000 qm

Gewächshäuser, Erdgeschoss Greenhouses, ground floor

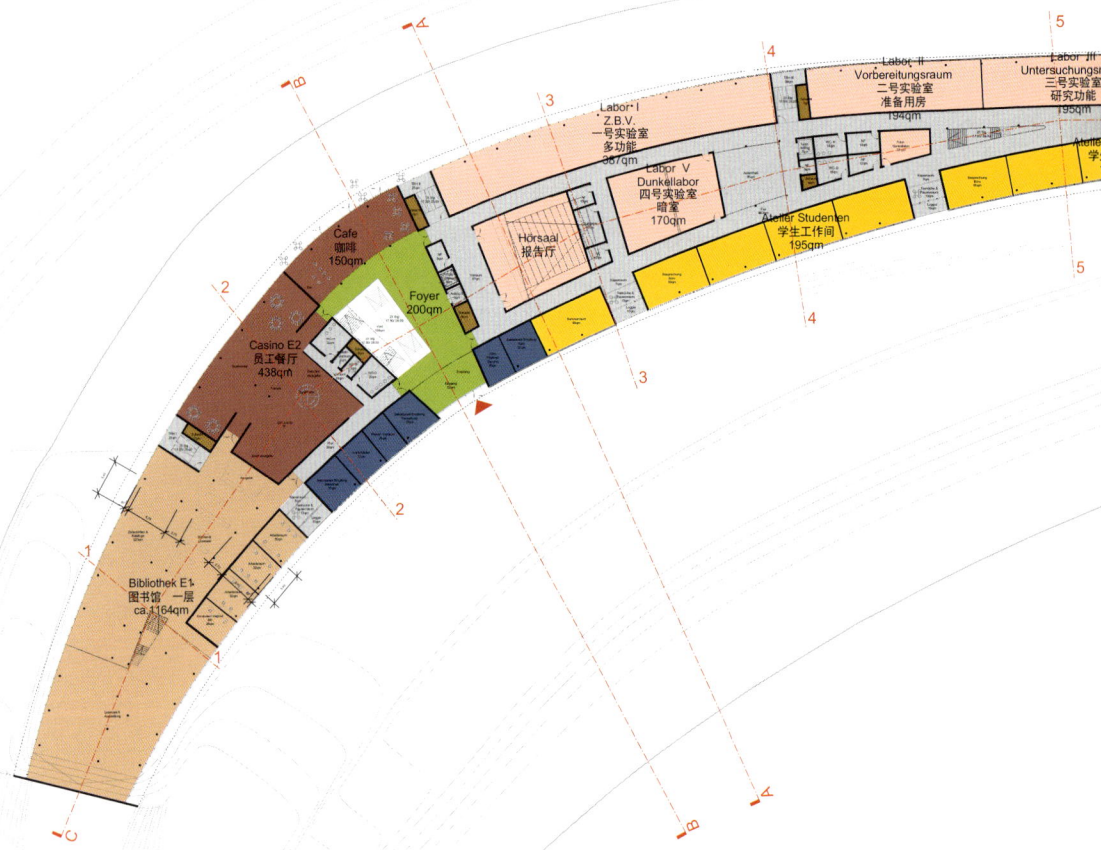

Labor III
Untersuchungsr
三号实验室
研究功能
195qm

Labor II
Vorbereitungsraum
二号实验室
准备用房
194qm

Labor I
Z.B.V.
一号实验室
多功能
867qm

Labor V
Dunkellabor
四号实验室
暗室
170qm

Atelier
学生

Atelier Studenten
学生工作间
195qm

Cafe
咖啡
150qm

Hörsaal
报告厅

Foyer
200qm

Casino E2
员工餐厅
438qm

Bibliothek E1
图书馆 一层
ca.1164qm

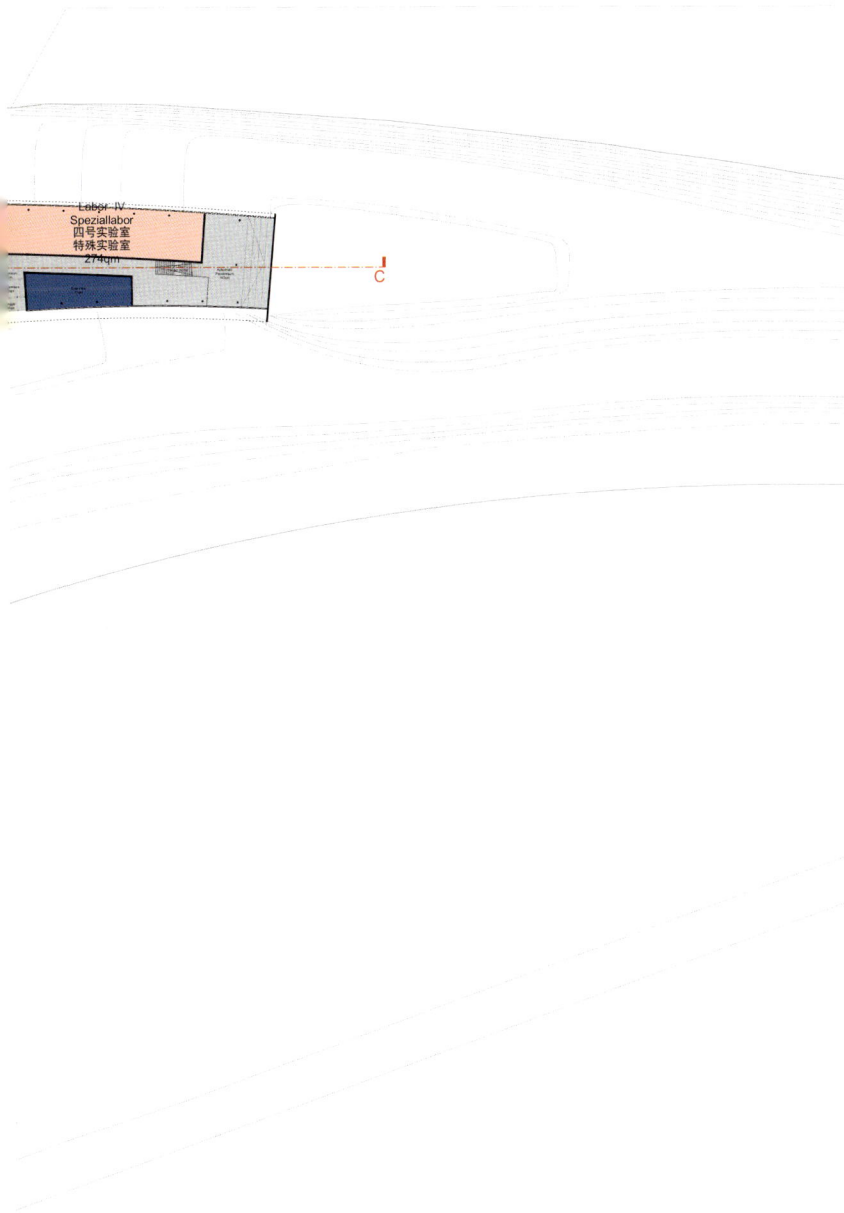

Labor IV
Speziallabor
四号实验室
特殊实验室
274qm

C

Den dritten baulichen Schwerpunkt bildet im Nordwesten das Forschungsgebäude mit Labors, Büroräumen und Herbarium. Auch dieses Gebäude ist in sanfter Krümmung eingebettet in den großen Ringwall. Nach Süden und Norden öffnen sich zur natürlichen Belichtung große Fensterbänder. Im Süden ist ein Gitterwerk mit Rankpflanzen vorgespannt, das gegen zu starke Sonneneinstrahlung abschirmt. Das Forschungsgebäude ist Teil des großen Ringes, es vermittelt räumlich und symbolisch zwischen der Welt der Wissenschaft im Forschungsgarten und der lebendigen Darstellung der Pflanzenwelt im Botanischen Garten. Neben diesen zentralen Gebäudekomplexen wird der Garten strukturiert durch Familien von wiederkehrenden kleineren Bauten, Restaurants, Teehäusern, Informationspavillons und Serviceeinrichtungen.

Bei der Konstruktion und der Wahl der Materialien für die Gebäude stehen Aspekte der Nachhaltigkeit, Langlebigkeit, Umweltverträglichkeit und Wirtschaftlichkeit im Vordergrund. Bevorzugt werden deshalb „pure" Materialien wie Beton, Glas, Stahl und Holz, für die Bauten im Garten auch Bambus. Energieeinsparung ist ein vorrangiges Ziel: durch intelligente Konzepte der aktiven und passiven Solarnutzung, durch weitgehend natürliche Be- und Entlüftung und die Nutzung der Potenziale des Grundwassers mittels Wärmepumpen.

Forschungsgebäude, 1. Obergeschoss Research building, first floor

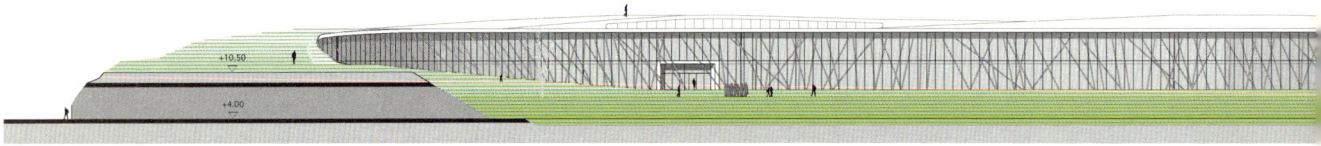

Forschungsgebäude, Ansicht Süd Research building, south elevation

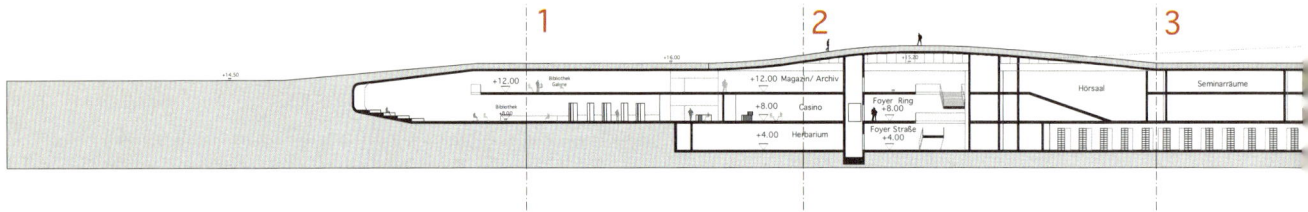

Forschungsgebäude, Längsschnitt Research building, longitudinal section

Forschungsgebäude, Querschnitte Research building, cross section

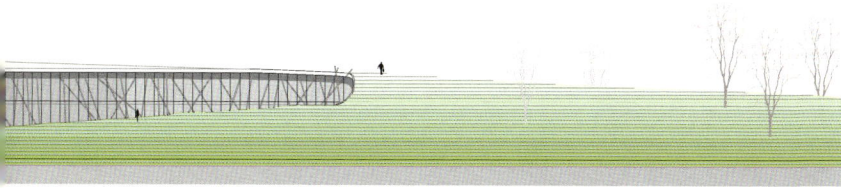

5

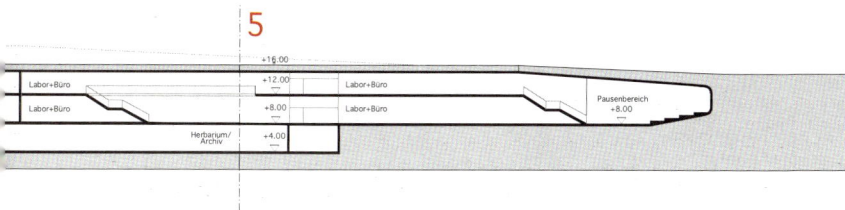

The third architectural focal point is found in the northwest of the park, the research building with laboratories, offices, and a herbarium. This gently curving building is also embedded in the large circular embankment. Large bands of windows on its north and south sides allow for ample natural light. On the south side a trellis with climbing plants provides protection from the sun. The research building is a part of the ring; it functions as a mediator between the world of science in the research garden and the lively depiction of the plant world in the botanic garden. In addition to these central complexes of buildings the garden also contains families of smaller recurring structures, restaurants, tee houses, information pavilions, and service facilities.

Regarding methods of construction and the choice of materials, aspects of sustainability, longevity, environmental compatibility, and economics are of the highest priority. "Pure" materials like concrete, glass, steel, and wood are preferred, and bamboo is used extensively for the structures in the garden. Saving energy is a major goal: by using intelligent concepts of active and passive solar energy, through a large degree of natural ventilation, and by using the potential of groundwater in heat pumps.

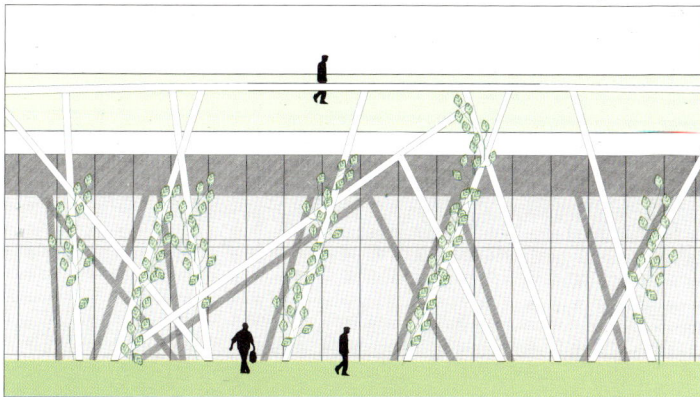

Forschungsgebäude, Fassadendetails Research building, façade details

Vogelperspektive des Botanischen Gartens, Wettbewerbsphase
Aerial perspective of the Botanic Garden, competition phase

Der Forschungsgarten Der Botanische Garten Chenshan wird insgesamt Forschungs- und Versuchgelände der neuen Universität werden. Die eigentlichen Forschungs- und Versuchsflächen aber liegen im Norden, außerhalb des Gartens. Die Einteilung in Felder und Wege folgt der Logik eines gärtnerischen Betriebes. Es entstehen Pflanzflächen für Bäume, Sträucher und Stauden. „Hier werden Pflanzen gesammelt, gesichtet, bezeichnet, akklimatisiert und analysiert. Die Belegung der Felder wird mit einfachen Spielregeln gesteuert, die bestimmte Entwicklungen stimulieren oder einschränken. Unterschiedliche Substrate, Feuchtigkeitsgrade und Expositionen erzeugen verschiedene gärtnerische Milieus." (Dietmar Straub)

Inmitten dieser Versuchsfelder liegt ein Gästehaus mit Konferenzräumen für den internationalen wissenschaftlichen Austausch.

Die Besucher des Gartens Es war erklärtes Ziel, die Entwicklung des Botanischen Gartens gleichrangig zu sehen mit seiner Gestaltung als Naherholungsraum für die Bevölkerung der Region und als touristisches Ziel für eine interessierte Öffentlichkeit weit darüber hinaus.

Zunächst ist der Botanische Garten ein lebendiges Museum. Es wird verschiedenste Lehr- und Studienangebote geben, für Kinder- und Schulklassen ebenso wie für die Erwachsenenbildung. Zugleich soll hier ein Ort entstehen, an dem die Menschen aus Shanghai sich am Wochenende erholen können. Es gibt Restaurants in verschiedenen Größen und Preisklassen, Wiesen, auf denen man sich ausruhen und picknicken kann, Ruderboote auf dem großen See, Fahrradverleih, den „Kids' Garden", ein Spielparadies und einen Campingplatz außerhalb des Ringes.

Gerechnet wird mit bis zu 30.000 Besuchern am Tag, die teils mit dem öffentlichen Nahverkehr, teils mit Autos und Bussen anreisen werden. Was zunächst überdimensioniert erscheint an Erschließung, Parkraum oder Gebäudedimensionierung, erklärt sich aus diesen Zahlen. Die Großzügigkeit und Vielfältigkeit des Raumes sollten beides erlauben: Erholung und Spaß für die große Zahl und Rückzug und Naturgenuss für den Einzelnen.

Ausblick Der Masterplan zum Botanischen Garten in Shanghai hat alle Hürden der Genehmigung genommen. Zur Weltausstellung im Jahr 2010 soll die Eröffnung des Parks gefeiert werden. Dann wird das große Rund der Bäume zu erleben sein, und ein Teil der Themengärten soll bis dahin realisiert werden.

Mit dem Bau wurde bereits begonnen, der Wall und der zentrale See sind in Umrissen zu erkennen. Die ersten Bäume wurden gepflanzt, mächtige Solitäre, Umsiedler, für die der Platz im alten Botanischen Garten zu eng geworden war.

Wie bei den meisten wichtigen Bauvorhaben in China liegt die Planung der Ausführung in chinesischer Hand. Trotz der weiter laufenden Beratung wird nicht alles genau so entstehen, wie es in Europa erdacht wurde, manches wird sich auf dem langen Weg der Realisierung verändern. Die Hoffnung ist, dass sich die markante Grundstruktur behaupten wird.

2010 wird der Garten noch lange nicht fertig sein. Es wird Jahrzehnte dauern, bis die exotischen Pflanzen aus aller Welt herbeigebracht werden können, und Jahrhunderte, bis die Lorbeerwälder der Erde auf dem großen Ring ihre volle Schönheit erreichen.

The Research Garden The Chenshan Botanic Garden will become the research and experimental station for the new university. The actual research and experimental areas, however, lie to the north, outside the garden. The arrangement in fields and paths follows the logic of a nursery. Areas for trees, shrubs, and perennials have been created. "Here plants are collected, culled, designated, acclimatised, and analysed. The allocation of fields is managed according to simple rules which stimulate or limit certain types of development. Different soils, degrees of moisture, and exposition result in different gardening environments." (Dietmar Straub)

Located in the midst of these experimental plots is a guesthouse with conference rooms intended to encourage an international scientific exchange.

The Garden's Visitors One of the stated aims was to give equal importance to the development and design of the Botanic Garden as a source of local recreation for residents of the region and as a goal for interested tourists from around the world.

First of all, the Botanic Garden is a living museum. It will offer a variety of opportunities for teaching and studying, for children and school classes as well as well as for adult education. At the same time, it should serve as a place where the people of Shanghai can relax at weekends. There are restaurants of different sizes and price classes, meadows where people can rest and have picnics, rowboats on the large lake, bicycles for hire, a garden for children, a playground, and a campground outside the ring.

It is estimated that there will be 30,000 visitors per day, some of whom will come with public transport,

others of whom will use cars or busses. The large circulation system, parking areas, and buildings, which appear to be oversized at first, are a consequence of this estimate. The large scale and variety of the spaces should allow for both: relaxation and fun for a large number of people and a haven and place to enjoy nature for individuals.

Looking Ahead The masterplan for the Botanic Garden in Shanghai has already successfully met all of the hurdles encountered in the approvals process. The opening of the park will be celebrated in 2010, in time for the World Expo. A great majority of the trees will be in place, and some of the theme gardens will be finished by then.

Construction has already begun, and the embankment and the central lake are recognizable. The first trees have been planted; large individual trees taken from the old Botanic Garden due to a lack of space there.

As is the case with most of the important construction projects in China, responsibility for the construction documentation is firmly in Chinese hands. Despite ongoing consultation everything will not be created as planned in Europe, some things will be altered on the long road to completion. Our hope is that the distinctive fundamental structure will prevail.

The garden will be far from finished in 2010. It will take decades to bring all of the exotic plants from around the world to Shanghai, and centuries until the world's laurel forests growing on the huge ring reach their full beauty.

ANHANG

Projektbeteiligte Project participants

Bauherr Client
Chenshan Botanical Garden Shanghai Project
Team, China

Wettbewerbsentwurf Competition design
Projektgruppe Chenshan Botanischer Garten
Shanghai

Entwurfsplanung Design documentation
Planungsgruppe Valentien

Landschaftsarchitektur Landscape architecture
Valentien + Valentien Landschaftsarchitekten und
Stadtplaner SRL
Straub + Thurmayr Landschaftsarchitekten
Susanne Brittinger I Dietmar Bühler I Yiju Ding I
Sandra Fick I Friederike Kühn I Julia Knop I
Ingrid Liebald I Johannes Niehoff I Guiping Peng I
Joachim Pogorzalek I Julia Romstätter I Ines
Siebrecht I Dietmar Straub I Anna Thurmayr I
Christoph Valentien I Dayana Valentien I Donata
Valentien I Yuan Xia

Architektur Architecture
Auer + Weber + Assoziierte
Fritz Auer I Moritz Auer I Haluk Cinar I Andreas
Dirlam I Yinliang Fan I Sebastian Gerlsbeck I Cynthia
Grieshofer I Bernhard Heidberg I Julian Krüger I
Ingo Pucci I Daniela Sacher I Stephan Suxdorf I Jan
Stecher I Fang Tian I Mohan Zeng

Koordination Deutschland – China
Germany – China coordination
Yiju Ding

Beratung Consulting
Botanik Vegetation: Prof. Dr. Jürke Grau, Bota-
nischer Garten München-Nymphenburg
Wasserkonzept Hydrological concept: Irene Burk-
hardt Landschaftsarchitekten, München; Taraske
Consult Ingenieurgesellschaft für Energie- und
Umwelttechnik mbH, Frankfurt am Main
Tragwerksplanung Structural planning: Schlaich
Bergermann und Partner, Stuttgart: Sven Plieninger,
Kai Kürschner, Rüdiger Weitzmann
Gebäudetechnik Building services engineering:
Ingenieurbüro Hausladen GmbH, München
Biogasanlage Biogas plant: BTA Biotechnische
Abfallverwertung GmbH & Co. KG, München
Gastronomiekonzept Gastronomic concept: Reisner
& Frank Ingenieure, München

Ausführungsplanung Landschaftsarchitek-
tur Final planning landscape architecture
Shanghai Landscape Architecture Design Institute

Ausführungsplanung Architektur
Final planning architecture
SIADR, Shanghai Institute of Architectural Design &
Research

Chronologie Chronology

Im Jahr 2005 wurde Prof. Christoph Valentien zur Teilnahme am Internationalen Wettbewerb Chenshan Botanischer Garten Shanghai eingeladen.

Der Wettbewerb wurde durch die „Projektgruppe Chenshan Botanischer Garten Shanghai" erarbeitet, bestehend aus den Büros:

Valentien + Valentien Landschaftsarchitekten und Stadtplaner

Straub + Thurmayr Landschaftsarchitekten

Auer + Weber + Assoziierte Architekten

Wettbewerbsabgabe: Ende Oktober 2005, 3. Preis

Überarbeitungsphase: November bis Ende Dezember 2005, 1. Rang

Im März 2006 wurde die Planungsgruppe Valentien als Arbeitsgemeinschaft der oben genannten Büros mit der Entwurfsplanung beauftragt.

Beginn der Bauarbeiten war im Frühjahr 2007.

Die Fertigstellung des Projekts ist für 2010 mit der Eröffnung der EXPO Shanghai geplant.

In 2005, Christoph Valentien was invited to take part in the International Competition Chenshan Botanic Garden Shanghai.

The entry for the competition was developed by the "Chenshan Botanic Garden Shanghai project group," consisting of the following practices:

Valentien + Valentien Landschaftsarchitekten und Stadtplaner

Straub + Thurmayr Landschaftsarchitekten

Auer + Weber + Assoziierte Architekten

The entry was turned in at the end of October 2005, and was awarded the 3rd prize.

The design was then revised from November until end of December 2005, and received 1st rank.

In March 2006, the Valentien planning group was commissioned with the design documentation in collaboration with the above-mentioned practices.

Building started in spring 2007.

The completion of the project is planned for 2010, in time with the opening of the EXPO Shanghai.

Autoren Authors

Christoph Valentien

Jahrgang 1939, Studium der Landschaftsarchitektur an der TU München und Städtebau an der RWTH Aachen. Seit 1971 gemeinsames Büro mit Donata Valentien. 1980–2002 Professor am Lehrstuhl für Landschaftsarchitektur und Entwerfen der TU München. Mitwirkung in verschiedenen kommunalen Beratungsgremien, Fachverbänden und Wettbewerbsjurys, Concurrent Professorship der Nanjing Forestry University.

Donata Valentien

Jahrgang 1944, Studium der Landschaftsarchitektur an der TU München und der TU Berlin. Honorarprofessuren an der Universität Stuttgart-Hohenheim und an der TU München. Mitwirkung in zahlreichen kommunalen und staatlichen Beratungsgremien, Beirat für Raumordnung und Städtebau der Bundesrepublik Deutschland, Lenkungsausschuss der IBA Emscher Park u. a. Amtierende Direktorin der Sektion Baukunst an der Akademie der Künste Berlin.

Hu Yong-Hong

Jahrgang 1968, Studium der Gartenbauwissenschaften an der Northeast Forestry University, Harbin, Heilongjiang und Studium der Landschaftsarchitektur an der Beijing Forestry University. Doktor der Gartenbauwissenschaften. Leitung zahlreicher Gartenausstellungen, Publikationen zu Pflanzensystemen, Gartenausstellungen und Botanischen Gärten. Direktor des Botanischen Gartens Shanghai. Zahlreiche Auszeichnungen und Preise, Gastprofessur an der Forstuniversität Peking.

Christoph Valentien

Born in 1939, he studied landscape architecture at the Technical University in Munich and urban planning at the RWTH Aachen University. Since 1971 has had a joint office with Donata Valentien. 1980–2002 professor of landscape architecture and design at the TU Munich. Member of numerous communal advisory councils, professional associations, and competition juries, concurrent professorship at the Nanjing Forestry University.

Donata Valentien

Born in 1944, she studied landscape architecture at the TU Munich and at the TU Berlin. Honorary professor at the University of Stuttgart-Hohenheim and at the TU Munich. Member of numerous communal and government advisory councils including the Federal Advisory Council for Regional and Urban Planning and the IBA Emscher Park Steering Committee. Acting Director of the Architecture Section at the Academy of the Arts in Berlin.

Hu Yong-Hong

Born in 1968, he studied horticulture at Northeast Forestry University, Harbin, Heilongjiang and landscape architecture at Beijing Forestry University. PhD in horticulture. He has directed numerous garden exhibitions and has published about plant systems, garden exhibitions, and botanic gardens. Director of the Shanghai Botanic Garden. Numerous awards and prizes, teaches as a guest professor at the Beijing Forestry University.

He Shanan

Jahrgang 1932, Studium der Agrarwissenschaften an der Landwirtschaftlichen Universität Zhejiang. Professur für Botanik am Institute of Botany, Jiangsu Province & Chinese Academy of Sciences von Nanjing. Präsident der International Association of Botanical Gardens, Ehrendirektor des Botanischen Gartens Nanjing und Mitglied der Chinesischen Akademie der Wissenschaften. Zahlreiche Publikationen und Auszeichnungen. Weltweite Tätigkeit als Vortragender und Delegierter Chinas in internationalen Kommissionen.

Thies Schröder

Jahrgang 1965, Studium der Landschaftsplanung an der TU Berlin. Fachjournalist, Redakteur und Autor im Bereich Landschaftsarchitektur, Städtebau und Regionalentwicklung. Zahlreiche Publikationen, Buchbeiträge, Vorträge und Aufsätze. Schröder leitet das Büro für Planungskommunikation ts redaktion.

He Shanan

Born in 1932, he studied agronomy at Zhejiang Agricultural University. Professor of botany at the Institute of Botany, Jiangsu Province & Chinese Academy of Sciences of Nanjing. President of the International Association of Botanical Gardens, honorary director of the Nanjing Botanic Garden, and member of the Chinese Academy of Sciences. He has published widely and received many awards and prizes. Active as a lecturer around the world, he serves as a Chinese delegate on international commissions.

Thies Schröder

Born in 1965, he studied landscape planning at the TU Berlin. Editor and author in the area of landcape architecture, urban planning and regional development. Numerous publications, essays and lectures. Schröder runs the Office for Planning Communication ts redaktion.

DATE DUE